versity of Reading

AF846144

90 0630875 9

IMAGE PROCESSING FOR THE FOOD INDUSTRY

SERIES IN MACHINE PERCEPTION AND ARTIFICIAL INTELLIGENCE*

Editors: **H. Bunke** (Univ. Bern, Switzerland)
P. S. P. Wang (Northeastern Univ., USA)

Vol. 19: Parallel Image Analysis: Theory and Applications
(Eds. *L. S. Davis, K. Inoue, M. Nivat, A. Rosenfeld and P. S. P. Wang*)

Vol. 20: Picture Interpretation: A Symbolic Approach
(Eds. *S. Dance, T. Caelli and Z.-Q. Lin*)

Vol. 21: Modelling and Planning for Sensor Based Intelligent Robot Systems
(Eds. *H. Bunke, T. Kanade and H. Noltemeier*)

Vol. 22: Machine Vision for Advanced Production
(Eds. *M. Pietikainen and L. F. Pau*)

Vol. 23: Machine Learning and Perception
(Eds. *G. Tasini, F. Esposito, V. Roberto and P. Zingaretti*)

Vol. 24: Spatial Computing: Issues in Vision, Multimedia and Visualization Technologies
(Eds. *T. Caelli, Peng Lam and H. Bunke*)

Vol. 25: Studies in Pattern Recognition Memorial Volume in Honor of K S Fu
(Eds. *H. Freeman*)

Vol. 26: Neural Network Training Using Genetic Algorithms
(Eds. *L. C. Jain, R. P. Johnson and A. F. J. van Rooij*)

Vol. 27: Intelligent Robots — Sensing, Modeling & Planning
(Eds. *B. Bolles, H. Bunke and H. Noltemeier*)

Vol. 28: Automatic Bankcheck Processing
(Eds. *S. Impedovo, P. S. P. Wang and H. Bunke*)

Vol. 29: Document Analysis II
(Eds. *J. J. Hull and S. Taylor*)

Vol. 30: Compensatory Genetic Fuzzy Neural Networks and Their Applications
(*Y.-Q. Zhang and A. Kandel*)

Vol. 31: Parallel Image Analysis: Tools and Models
(Eds. *S. Miguet, A. Montanvert and P, S. P. Wang*)

Vol. 33: Advances in Oriental Document Analysis and Recognition Techniques
(Eds. *S.-W. Lee, Y. Y. Tang and P. S. P. Wang*)

Vol. 34: Advances in Handwriting Recognition
(Ed. *S.-W. Lee*)

Vol. 35: Vision Interface — Real World Applications of Computer Vision
(Eds. *M. Cheriet and Y.-H. Yang*)

Vol. 36: Wavelet Theory and Its Application to Pattern Recognition
(*Y. Y. Tang, L. H. Yang, J. Liu and H. Ma*)

*For the complete list of titles in this series, please write to the Publisher.

Series in Machine Perception and Artificial Intelligence – Vol. 37

IMAGE PROCESSING FOR THE FOOD INDUSTRY

E R DAVIES

Professor of Machine Vision
Department of Physics
Royal Holloway, University of London
Egham, Surrey TW20 OEX UK

World Scientific
Singapore • New Jersey • London • Hong Kong

Published by

World Scientific Publishing Co. Pte. Ltd.

P O Box 128, Farrer Road, Singapore 912805

USA office: Suite 1B, 1060 Main Street, River Edge, NJ 07661

UK office: 57 Shelton Street, Covent Garden, London WC2H 9HE

British Library Cataloguing-in-Publication Data
A catalogue record for this book is available from the British Library.

P.

Dr. S. Henson

IMAGE PROCESSING FOR THE FOOD INDUSTRY
Series in Machine Perception and Artificial Intelligence — Vol. 37

Copyright © 2000 by World Scientific Publishing Co. Pte. Ltd.

All rights reserved. This book, or parts thereof, may not be reproduced in any form or by any means, electronic or mechanical, including photocopying, recording or any information storage and retrieval system now known or to be invented, without written permission from the Publisher.

For photocopying of material in this volume, please pay a copying fee through the Copyright Clearance Center, Inc., 222 Rosewood Drive, Danvers, MA 01923, USA. In this case permission to photocopy is not required from the publisher.

ISBN 981-02-4022-8

THE UNIVERSITY OF READING LIBRARY

Printed in Singapore.

Contents

Part 2 Application to food production

Preface

Over the past thirty years, image processing and machine vision have achieved wide utility in many facets of our lives, ranging from films and television, document processing, space science and robot control to a host of commercial, forensic, industrial, medical and military applications. Of the industrial applications, those in the food industry are especially important, as food processing is undergoing vast changes to yield products of ever higher added value—while at the same time the possibility of harm though bacteria and microtoxins is ever present, and a great many additives and contaminants can be regarded as undesirable or unattractive for human consumption. As a result, careful control of the manufacturing process is called for and high standards of quality control are in increasing demand. Digital images are key to the possibilities for attaining such control, as a single compact sensor can provide huge amounts of crucial information at extremely high rates. However, image processing and analysis are specialist subjects, and their application is complicated by the sheer volume of data that has to be processed—and by the rate at which it must be processed if real-time production speeds are to be matched.

This monograph aims to explain the design of image processing systems for the food industry. Because of the specialist nature of image processing on the one hand and the demands of food production on the other, this is no easy task. After careful consideration it seemed appropriate to divide the volume into two main parts, the first dealing with image processing and vision methodology, and the second covering its application to food production. This partition reflects the main aims of the book, which are:

- to provide food practitioners with a clear, succinct background on relevant image processing technology;
- to provide image processing researchers with concrete ideas on how their subject can be applied to food production;
- to update researchers working on the application of image processing to food production; not least, to provide them with a useful set of references, and to show their relevance;

- to inform managers in the food industry of the current and potential applications of image processing to food production, and to demonstrate important issues that need to be taken into account during any implementation.

Of course, in a short monograph it is impossible to cover every possible topic at a uniform and substantial depth. Indeed, it will only be possible to describe some of the most important areas of image processing, describing them in sufficient detail to convey an understanding of their application to food production, but not necessarily to cover them in the same detail as advanced image processing textbooks; and to describe some of the main areas of food production to which image processing has been or potentially could be applied, but not necessarily to cover them in sufficient detail for an image processing expert to design systems without consulting food production engineers.

Generally, the book is written at the practitioner level rather than as a textbook. It is intended to be readable without access to other texts or published papers, though to get full value from it when dealing with new applications, more specialist works may have to be referred to. In this respect the case study form of presentation that is used in the later chapters should be of particular help to readers. University researchers in relevant areas (image processing, food production, etc.) should also benefit by consulting this volume.

This book is based on many years experience of working both on the theory of image analysis and on its practical aspects in the general area of automated visual inspection, particularly for the food industry. Needless to say it is biased by my own inclinations and by the specific problems I have encountered. However, I have often observed that many techniques retain their value from one application to another, so the particular case studies and ideas presented here have a fair generic content and should thus be useful to a wide range of practitioners in this subject area.

Finally, this monograph owes much to all the people I have worked with over a period of well over twenty years—postgraduate students, university colleagues and collaborators in the food industry. In particular, I would like to express my gratitude to Simon Barker, Daniel Celano, Barry Cook, Mark Edmonds, Darrell Greenhill and Zahid Hussain, all of whom have helped to hone my view of the subject to no small degree; to Adrian Johnstone and Piers Plummer, for providing very special help and expertise in the design of real-time vision hardware; and to Mark Bateman, Ian Hannah, Dave Mason and Dev Patel, for vital assistance with the design of key food inspection algorithms. In addition, it is a special pleasure to recall many rewarding discussions with John Chambers and Chris Ridgway of the Central Science Laboratory, York, who introduced me to the fascinating problems of the cereals industry. Without the help of all these colleagues and of course, the endless patience and moral support of my wife and family, this book would never have been written.

E.R. Davies
Royal Holloway, University of London

About the Author

Roy Davies graduated from Oxford University in 1963 with a First Class Honours degree in Physics. After 12 years' research in solid state physics, his interests in electronic data processing led him to Royal Holloway, University of London, where he is now Professor of Machine Vision. He has worked on the theory of image processing and on many core techniques for feature location and robust pattern matching, and has had considerable experience of algorithm design for automated visual inspection, particularly for the food industry. He has published two books—*Electronics, Noise and Signal Recovery* (1993) and *Machine Vision: Theory, Algorithms, Practicalities* (1997)—and more than 180 research papers, recently being awarded a Doctor of Science by the University of London for his work on the science of measurement.

Professor Davies is a regular reviewer for many image processing journals, and is on the editorial boards of *Imaging Science*, *Pattern Recognition Letters*, and *Real-Time Imaging*. He is a Fellow of the Institute of Physics and the Institution of Electrical Engineers, and a Senior Member of the IEEE. He has taught courses on pattern recognition, image processing and machine vision for over twenty years.

This book is dedicated to my family:

—to my late father, Arthur Granville Davies, for the intellectual inspiration he long ago instilled into me
—to my late mother, Mary Davies, for her never to be forgotten love and devotion
—to my wife, Joan, for love and support—and for providing the happy home life that has been the springboard of my research
—to my children, Elizabeth, Sarah and Marion, for love, music, and the thrills to keep me young at heart

Acknowledgements

The author would like to credit the following sources for permission to reproduce material from his earlier publications:

Academic Press for permission to reprint portions of the following paper as text in Chapter 11 and as Table 11.3:

Davies, E.R., Patel, D. and Johnstone, A.I.C. (1995). *Real-Time Imaging* **1**, no. 6, 397–407

Elsevier Science for permission to reprint figures in the following papers as Figures 9.2, 9.5, 9.6, 11.1:

Davies, E.R. (1987). *Pattern Recogn. Lett.* **6**, no. 5, 323–333

Davies, E.R. (1992). *Pattern Recogn. Lett.* **13**, no. 2, 113–121

Hannah, I., Patel, D. and Davies, E.R. (1995). *Pattern Recogn.* **28**, no. 8, 1135–1143

EURASIP for permission to reprint portions of the following papers as text in Chapter 10 and as Figures 10.3, 10.5–10.10:

Davies, E.R. (1998). *Proc. European Signal and Image Processing Conf.* (EUSIPCO'98), Rhodes, Greece, 8–11 Sept., pp. 589–592

Davies, E.R., Mason, D.R., Bateman, M., Chambers, J. and Ridgway, C. (1998). *Proc. European Signal and Image Processing Conf.* (EUSIPCO'98), Rhodes, Greece, 8–11 Sept., pp. 2561–2564

IEE for permission to reprint portions of the following papers as text in Chapters 8, 10, 14, as Tables 11.2, 14.1, and as Figures 8.1–8.6, 10.1, 10.2:

Davies, E.R. (1997). IEE Digest no. 1997/041, Colloquium on *Industrial Inspection*, IEE (10 Feb.), pp. 6/1–5

Davies, E.R. (1997). *Proc. 6th IEE Int. Conf. on Image Processing and its Applications*, Dublin (14–17 July), IEE Conf. Publication no. 443, pp. 161–165

Davies, E.R., Bateman, M., Chambers, J. and Ridgway, C. (1998). IEE Digest no. 1998/284, Colloquium on *Non-Linear Signal and Image Processing*, IEE (22 May), pp. 12/1–5

Davies, E.R. and Patel, D. (1995). IEE Digest no. 1995/113, Colloquium on *The Application of Machine Vision*, IEE (24 May), pp. 2/1–6

IFS International Ltd for permission to reprint a figure in the following paper as Figure 4.4:

Davies, E.R. (1984). In Pugh, A. (ed.), *Proc. 4th Conf. on Robot Vision and Sensory Controls*, London (9–11 Oct.), pp. 437–446

MCB University Press for permission to reprint a plate in the following paper as Figure 11.3:

Patel, D., Davies, E.R. and Hannah, I. (1995). *Sensor Review* **15**, no. 2, 27–28

The Organizing Committee of M^2VIP'95 for permission to reprint portions of the following paper as text in Chapters 1, 9, 11 and as Figure 9.1:

Davies, E.R. (1995). *Proc. 2nd Int. Conf. on Mechatronics and Machine Vision in Practice*, Hong Kong (12–14 Sept.), pp. 15–24

The help of Susan May in preparing the photographs for Figure 11.3 is gratefully acknowledged.

Finally, the author is grateful to J. Chambers and C. Ridgway of Central Science Laboratory, York, UK for the original images used to test the high speed object location algorithms (Section 10.4).

Glossary of Abbreviations

ANN	artificial neural network
ASAE	American Society of Agricultural Engineers
CAM	computer-aided manufacture
CCD	charge-coupled device
DA	discriminant analysis
DF	discriminant function
DSP	digital signal processor
EC	European Community
EPROM	erasable programmable read-only memory
EU	European Union
EVM	extraneous vegetable matter
FPP	full perspective projection
HSI	hue saturation intensity
HT	Hough transform
IEE	Institution of Electrical Engineers (UK)
IEEE	Institute of Electrical and Electronic Engineers (USA)
k-NN	k-nearest neighbour
MAFF	Ministry of Agriculture Fisheries and Food (UK)
MLP	multi-layer perceptron
MVA	multivariate analysis
NIR	near infra-red
NMR	nuclear magnetic resonance
NN	nearest neighbour
OCR	optical character recognition
PCA	principal components analysis
RGB	red green blue
ROC	receiver operator characteristic
SIMD	single instruction stream, multiple data stream
SME	Society of Motor Engineers (USA)
SPIE	Society of Photo-optical Instrumentation Engineers (USA)
SPR	statistical pattern recognition

US	United States
USA	United States of America
USDA	United States Department of Agriculture
WPP	weak perspective projection
1D	one dimension/one-dimensional
2D	two dimensions/two-dimensional
3D	three dimensions/three-dimensional

Chapter 1
Introduction

1.1 Food and its production

In the wild, animals spend a large proportion of their waking hours gathering food. Early man was no different in this respect, but over the millennia he learnt to cultivate the land and to keep animals, thereby saving himself much drudgery and permitting his life to become more comfortable and cultured. Nevertheless, his vital necessities (which are all the more poignant for the poor) include food, water, clothing, shelter and warmth (Table 1.1). Of course, warmth is provided by food, clothing and shelter, and so is not totally separate; on the other hand, food not only helps to induce warmth, but also is vital for promoting growth and repair of body tissues, so it is of supreme importance to the human animal.

An interesting aspect of food is its variety. The raw materials available to early man included fruit, nuts, seeds, vegetation, various sources of meat, fish and eggs (Table 1.2): this reflects his hunter-gatherer existence, which made him opportunistic in what he ate, but also meant that he must often have gone hungry. This precarious existence changed as he learnt how to cultivate the land and to herd and breed animals, though it may not at that stage have increased the variety of food: quite possibly the contrary applied, as cultivation would have brought with it a degree of specialization. However, in the end trade has brought about a great increase in the variety of foods that are available today. The short list in Table 1.2 has by now expanded to include all the raw foods in Table 1.3 and many more.

Seeds deserve a few extra words of explanation. Many items of food appear under this category, including pulses and cereal grains[1]. Seeds are the germinating components of vegetable matter, and the part that is normally eaten is known as a 'kernel'—a diminutive form of the word corn

[1] The word 'grain' often refers to non-food items such as sand, though it also refers to sugar and salt; and the term 'cereal' is ambiguous as it tends to be understood to mean breakfast cereals.

(cf. the Old English word 'cyrnel'). Similarly, the soft edible parts of nuts are known as kernels. Finally, the (often inedible) germinating components of fruit are also known as seeds.

The list in Table 1.3 is clearly not the end of the story. From the earliest of times, man has processed his foods: for example, he ground his corn to make bread, and pulverized roots and shoots to make them digestible. Nowadays, we have extended the concept to combine grinding, cooking (often in multiple stages), mixing of ingredients, compacting, extruding, freezing, packaging, and so on—thereby producing more palatable foods (Table 1.4). There has been a significant trend to appropriately packaged convenience foods. These processed foods are often surprisingly cheap, and they have the advantage that they can be marketed quickly and consumed while they are still fresh.

In addition, there is a long-term trend to luxury foods with high added value (Table 1.5). In fact, in many cases it is difficult to distinguish these from processed foods: rather, processed and luxury foods are two extremes on a continuous scale, with low added value at the one end and expensive ingredients for special occasions at the other. We shall not dwell on this point here.

Finally, there are various packaging modes (Table 1.6), including packets, packs, boxes, bottles, and cans. These few categories are misleadingly simple, as many foods in supermarkets have multiple packaging: e.g. packets within packs, within boxes, within larger boxes. This apparent wastage may well be justified in terms of cleanliness, convenience, ease of transportation, ease of quality control, and so on. However, it does reflect the complexity of the technology that is now being applied to the production and distribution of food.

Table 1.1 Necessities of life

food
water
clothing
shelter
warmth

Table 1.2 Raw foods

fruit
nuts
seeds
vegetables
mushrooms
animals/meat
game
poultry
fish
eggs
milk

Table 1.3 Varieties of raw foodstuffs

Fruit

- apples
- bananas
- blackcurrents
- gooseberries
- grapes
- peaches
- pears
- plums
- raspberries
- strawberries
- tomatoes

Seeds

- barley
- maize
- oats
- pulses (peas, beans, lentils, ...)
- rice
- rye
- sunflower
- sweetcorn
- wheat

Vegetables

- aubergines
- broccoli
- cabbages
- carrots
- cauliflowers
- lettuces
- parsnips
- peppers
- potatoes
- spinaches
- turnips

Animals

- cows
- frogs
- goats
- pigs
- sheep
- snails

Game and poultry

- chickens
- ducks
- partridges
- pheasants
- quails
- turkeys

Fish

- bream
- brislings
- cod
- crabs
- eels
- herrings
- lobsters
- oysters
- perch
- plaice
- prawns
- salmon
- sardines
- squid
- swordfish
- trout
- tuna
- whales

Table 1.4 Processed foods

biscuits	cheeses
bread	frozen vegetables
breadcrumbs	jams
breakfast cereals	pizzas
butter	yoghurt
cakes	

Table 1.5 Luxury foods

chocolate
confectionery
cream
crisps
decorated cakes
drinks
puddings

Table 1.6 Packaging

bottles
boxes
cans
cartons
packets
packs
sachets
tetrapacks
tubes

1.1.1 Quality control and other issues

Although the technology for food production and processing has developed markedly over the past 50 years or so, it has not undergone the phenomenal explosion experienced, for example, by the electronics and computer industry. In part this must be because the profit margins in food marketing are quite low, and partly because the mechanical handling and packaging machinery needed for food processing is relatively complex. However, there are now pressures to step up the quality of food as it reaches the consumer. One reason for this is the world-wide competition to market all sorts of goods, so guaranteed quality is a vital consideration. Another related consideration is the minimum guaranteed standards now being laid down by the EU and other international bodies. These standards are periodically augmented: as soon as new technology becomes available for checking quality of goods, it tends to be seen as vital to use it so that consumers can immediately benefit. Thus the law and the technology advance in alternate steps for the benefit of the consumer (though it is the latter who ultimately has to pay for these advances).

In fact, there are important reasons why quality has to advance. These are related to the possibility of poisonous or bacterial contamination, which can potentially (and sometimes in practice) lead to death. Another is the possibility of contaminants which are likely to be physically harmful, such as glass which is accidentally included or even materials such as razor blades or needles which are maliciously inserted. Other contaminants that are of concern are unapproved artificial colouring or flavouring additives. EU foodstuffs now have to state on the packaging what ingredients and additives have been used, and the technology for checking the contents automatically will have to be developed further. Then there is the possibility that the food itself will have deteriorated—perhaps as a result of inadvertent unfreezing—and automatic means for checking this will have to be developed (currently, consumers have to be watchful and need to examine 'sell-by' dates carefully). There is clearly much scope for intelligent packaging which records the temperature history of the material and alerts the consumer automatically.

Overall, there is considerable scope for development of food technology, not only in production, but also in quality control, and in packaging. Much has been done in the area of computer aided manufacture (CAM) and flexible automation (which would allow the same manufacturing line to be used economically for short runs of a variety of high value-added products). By and large, intelligent packaging is still wide open for future development, and the possibilities have not yet been fully fathomed, though the technology may well exist which would allow useful levels of implementation to be achieved quite quickly. However, it seems that quality control is the area which is lagging behind most. In this monograph we argue that it can be taken much further with the aid of image processing and machine vision techniques. We turn to this aspect in the following section.

1.2 Image processing and machine vision

Over the past thirty there has been much interest in the automatic processing and analysis of digital images, and many valuable techniques have been developed. Much of the running has been made by the need to rapidly process the enormous quantities of image data obtained from satellites, though medical and commercial (e.g. OCR) applications have also been important. Simultaneous with these activities have been efforts to understand and emulate the workings of the human visual system, and this has led to the subject of computer vision. However, computer vision does not aim to understand biological vision in detail, but rather to build up a knowledge of what is involved in seeing, by finding what computational constructs are required if visual perception is to occur. With the knowledge of the possible computational constructs, neurophysiologists will be better equipped to unravel the workings of the eye-brain system, and great strides have been made in this direction.

Machine vision is distinct from computer vision in that it aims to make machines process images from the real world, thereby enabling them to perform certain necessary tasks—i.e. it is task orientated rather than understanding orientated. We are clearly some way away from the universal machine vision system which can perform all the tasks that might be required of a machine, so machine vision tends to be orientated to the solution of specific tasks in specific environments. In fact, such tasks can be exacting and complex: they should not be taken as parochial or trivial.

In manufacturing environments there are a number of general functions to be performed by machine vision systems. These include:

- control of robots performing assembly operations
- guidance of lasers during cutting, milling or welding operations
- inspection of products during manufacture
- feedback to control manufacturing processes
- general process monitoring
- guidance of vehicles in factories

and many more. Broadly, these tasks can be classified into two main categories—guidance and inspection, which respectively involve mainly active and mainly passive observation of manufacturing processes. To further distinguish the two categories, we can say that if an active observation process is curtailed, the whole line will stop functioning, whereas if a passive observation process is curtailed, the line can go on functioning, but quality will eventually suffer.

In spite of the apparent clarity of these definitions, there are certain machine vision functions which fall between the two categories. This situation occurs when machine vision mediates by providing feedback which can control plant on a not too short timescale, and it can be assumed for several minutes that everything is working normally or according to an already extant trend. For example, if the spread of pancake mixture on a conveyor is observed by a vision system and used to control the temperature of the mixture, then temporary lack of visual information can perhaps be tolerated and it can be assumed that the temperature has not drifted. Similarly, if two long straight steel girders are being welded along their lengths, then temporary lack of guidance information from vision may not require that the welding process be immediately curtailed. The latter process—vision-guided welding—would nevertheless be classed as an active process, whereas the essentially similar temperature controlling process might be regarded as nearly passive. Thus inspection has a number of functions which range from purely passive to at least partially active, in one sense or another:

- feedback of information about items on product lines, to control plant
- feedforward of information to reject defective items from product lines

- logging of information to record numbers and characteristics of items on product lines
- general process monitoring to confirm that activity in a manufacturing cell is proceeding normally.

Another factor that is important when employing machine vision systems is the required speed of response. In many manufacturing situations, products are being made at rates of at least 20 per second, and the chance of a vision system working fast enough to control the fabrication of each of them individually is rather small: here machine vision will almost certainly have to take the passive role of making specific measurements and pronouncing on quality rather than controlling the manufacture or manipulation of each item. For example, if integrated circuits are to be fed to a printed circuit board, the flow of components will be so high that vision will be unable to arrange the individual placements: all it will be able to do is to check (a) that the printed circuit board is in exactly the correct position to receive components, (b) that the next component is ready to be inserted, (c) that it has been inserted, and (d) that it has been satisfactorily soldered in place. On the other hand, if robot assembly of a television set from pre-constructed modules is being contemplated, then machine vision can take an integral role, guiding each module into place with high precision. Curiously, the active role will not be the only one carried out by the vision system: the presence, and cost, of the system may only be justified if it is also checking the identity of, and possible damage to, each module and the general appearance of the whole assembly. Indeed, passive inspection checks can take quite a high proportion of its operating time, since waiting for modules to arrive and for the robot to move can take several seconds.

By now it will be clear that even for the active observation processes, the function of the vision system will be more constrained than for a human doing a similar task, and certainly far more constrained than for humans carrying out the highly variegated visual tasks they encounter in everyday environments. These constraints are typical of machine vision applications. In addition, the differences between active and passive machine vision functions are not great, and may well be purely academic: this applies especially to the inspection of modules and their insertion into television sets. As we shall see in subsequent chapters, this is not surprising as very much the same basic image processing functions, such as edge and corner detection, and object orientation, are involved in both types of process.

1.3 Biological versus machine vision

The discussion in the previous section has indicated that machine vision is currently more restricted than human vision. However, this prognosis is in some ways unfair, as great progress has recently been made in understanding how to emulate the human visual system viewing real scenes. There are three problems impeding the wider application of machine vision: one is the expense of making machine vision systems operate in real time on real data; another is making them work unaided in complex real-world situations; and the third is providing sufficient advance proof that they will be able to work unaided in new complex situations. To solve these problems, computers will have to have databases of knowledge that match those of human operators, and in many cases such databases can probably only be acquired from a lifetime's experience walking around and living as a human being. These factors conspire against handing over complete control of complex real-world situations to robots and machine vision systems.

Fortunately, many tasks which assembly automata could tackle are quite restricted. The adage that on a widget line you only find widgets or their components is a useful one: designing a constrained system to handle widgets and their components (or alternatively, chocolates and their fillings) is something that is manageable, and indeed is something that can be handled completely successfully using present day vision technology.

There are two further aspects to be kept in sight. First, if a human can perform the task, then it shows that the available images contain all the necessary information, and that it should be possible for a machine vision system to analyse the situation and control the robot automatically. Second, if the task is repetitive and has to be carried out hour after hour when a human operator would quickly become tired, bored and unreliable, then it is appropriate to replace the human by a robot system. Humans are valuable when skilled non-repetitive tasks involving a lot of experience have to be carried out, whereas robots are valuable when the opposite is the case. In addition, robots are likely to be preferable when highly accurate consistent measurements have to be made against predetermined standards. This means that, in quality control situations such as we expect from now on in the food marketing arena, it will be essential to employ robot sensing systems to achieve the necessary consistency and accuracy.

1.4 How can image processing help with food processing?

We started off this chapter with a discussion about the great variety of foods that now have to be produced and marketed, and how one of the major outstanding problems that will have to be solved is control of quality. Then in the next two sections we showed that image processing and machine vision are available to help with these tasks; in addition, we demonstrated that they have potential advantages relative to use of human operators, at least in the areas of repetitive tasks and those demanding high degrees of accuracy and consistency. Thus image processing and machine vision are likely to score highly for quality control purposes. However, the principles and technology of machine vision for quality control are almost identical to those required for controlling robots and for guiding robot vehicles. This means that it will be possible to use virtually identical techniques for a great many purposes of current relevance in the food industry. Such purposes include:

- guiding fruit picking machines
- guiding pruning machines
- guiding crop spraying machines
- tracking and sizing animals
- controlling evisceration machines
- checking products for size and shape
- checking icing patterns on cakes
- inspecting products for appearance
- controlling packing machines

and many similar cases. These fall into the general categories of inspection, handling, control and guidance. Although, as noted above, the vision aspects of these categories of activity are almost identical, their realization in terms of robot, vehicle or machine actions will be quite varied. Such mechanical and manipulative operations are beyond the scope of this monograph. It will be as much as we can do within the confines of these pages to explore the principles by which image processing can be applied to the benefit of the food industry: we leave it to other books to demonstrate how the mechanization of these tasks can be achieved. However, by and large mechanization of this sort is known technology. It is vision that is new and not widely understood. And it is vision technology that needs to be developed further in order for food and other applications to be applied to the human good. It will be the aim of this monograph:

- to provide food practitioners with a clear, succinct background on relevant image processing technology;
- to provide image processing researchers with concrete ideas on how their subject can be applied to food production;
- to update researchers working on the application of image processing to food production; not least, to provide them with a useful set of references, and to show their relevance;
- to inform managers in the food industry of the current and potential applications of image processing to food production, and to demonstrate important issues that need to be taken into account during any implementation.

1.5 The following chapters

The remainder of this monograph is divided into two main parts. Part 1 is concerned with the basics of image processing and machine vision: this part aims to provide generic information which forms the basis for use in practical systems. Part 2 is concerned with the study of food applications and aims to show how image processing and machine vision are used to solve real problems. As a general rule, if a technique is used for several different food applications, it will be covered in Part 1, whereas if it is specific to a particular food application, it will be covered in Part 2 with a description of the application.

A monograph of this type cannot cover everything and has to be selective. Some topics cannot be dealt with in any detail. Indeed, some topics are so involved that they demand a book of their own, and it is not appropriate to cover them in any depth here. In such cases, the topic is dealt with by cross-referencing to other topics or to the earlier literature, or by covering the image processing principles in reasonable detail in Part 1.

Part 1

Image processing methodology

This part of the book describes generic image processing methodology which will be relevant for a wide variety of applications in the food industry. It provides essential background material for understanding the food processing case studies described in Part 2.

Chapter 2
Images and image processing

2.1 Introduction

The purpose of this chapter is to show how images can be handled and processed by computer, and to lead on to more complex operations which will provide the basic tools for recognizing objects. Quite a lot will have to be covered within the confines of the chapter, and space limitations will prevent the subject from being covered exhaustively. Thus the storyline that will be developed will have to encompass not only the basic methods and the detailed programming notation required to represent them but also deeper knowledge about the strategies needed to obtain information from images. We start by describing the nature of image data, and then move on to consider some simple image processing operations.

2.2 Images

Although humans see in colour, the majority of image processing systems used in the food industry and elsewhere see in 'black and white' (i.e. shades of grey). However, the situation is changing rapidly, as colour frame-grabbing boards for computers are dropping to very low prices, and most PCs sold nowadays have colour monitors. Nevertheless, the camera needed to feed a colour image to a frame-grabber may be up to 10 times as expensive as a black and white camera of the same resolution.

While humans have come to rely on colour (evolution of the colour sense may well have arisen through the need to tell when fruit is ripe and to recognize which of a variety of other possible foods are poisonous or rotten), it is relatively rare for colour systems to be crucial for distinguishing between relevant objects. This is advantageous, since usually the additional colour channels that are needed require quite complex processing before full use can be made of the extra information they provide. Perhaps more important, real applications require real-time processing, and the additional processing demanded by colour is often more than a factor three greater than for a black and white channel, as quite complex methods may have to be used to combine the sets of data: in practice, such processing may not be warranted because of the cost of the additional hardware needed to achieve real-time operation. Hence a good proportion of the algorithms described in this monograph will relate to black and white images. Nevertheless, colour does have its place in the scheme of things (not least for fruit inspection applications), as we shall see in Chapters 12 and 13.

An obvious idea for quickly interpreting colour images is to add the outputs of the various colour channels with suitably chosen weighting factors, and to use the resulting grey-scale images as sole discriminators between coloured objects. Often the channel chosen is the hue channel (see Section 12.5.2), but this need not be so. However, if such a short-cut is taken the result will be similar to putting a colour filter in front of a black and white camera, and the quite significant difference in cost may not be justified. In fact, colour is a specialist topic, and very much depends on the data being addressed in the specific application. Hence we defer further detailed discussion of the topic to the case studies in Part 2.

To summarize the situation: colour vision involves use of red, green and blue (RGB) colour channels, each of which is essentially equivalent to the grey-scale channel used for black and white images. A typical digitized grey-scale image will have a resolution of 256 × 256 or 512 × 512 *pixels* (picture cells), and the grey-scale resolution will normally be 0 to 255: this can conveniently be stored as a single byte (8 bits) of information. Figure 2.1(a) shows a typical grey-scale image, and Figures 2.1(b–d) show the effect of zooming in until the individual pixels can be seen clearly. Notice how global recognition is progressively impeded by the interfering effects of spatial quantization and the presence of high levels of speckle noise on the egg.

2.3 Image processing

As normally defined, image processing is about the conversion of one image into another. Sometimes this is carried out to improve the image in some way—as happens during the removal of noise—and sometimes it is carried out as part of the process for analysing the image, e.g. to find what objects it contains.

2.3.1 Pixel operations

Probably the simplest image processing operations are so-called *pixel operations*, in which a new image is produced by processing each pixel in some way, and the result is placed at the corresponding location in the new image: usually the same processing technique is employed at each pixel location. For example, we could double the intensity at each pixel location in order to make the final image lighter. This would be achieved by a simple routine of the form:

```
for each pixel in image do
    double intensity value;
```

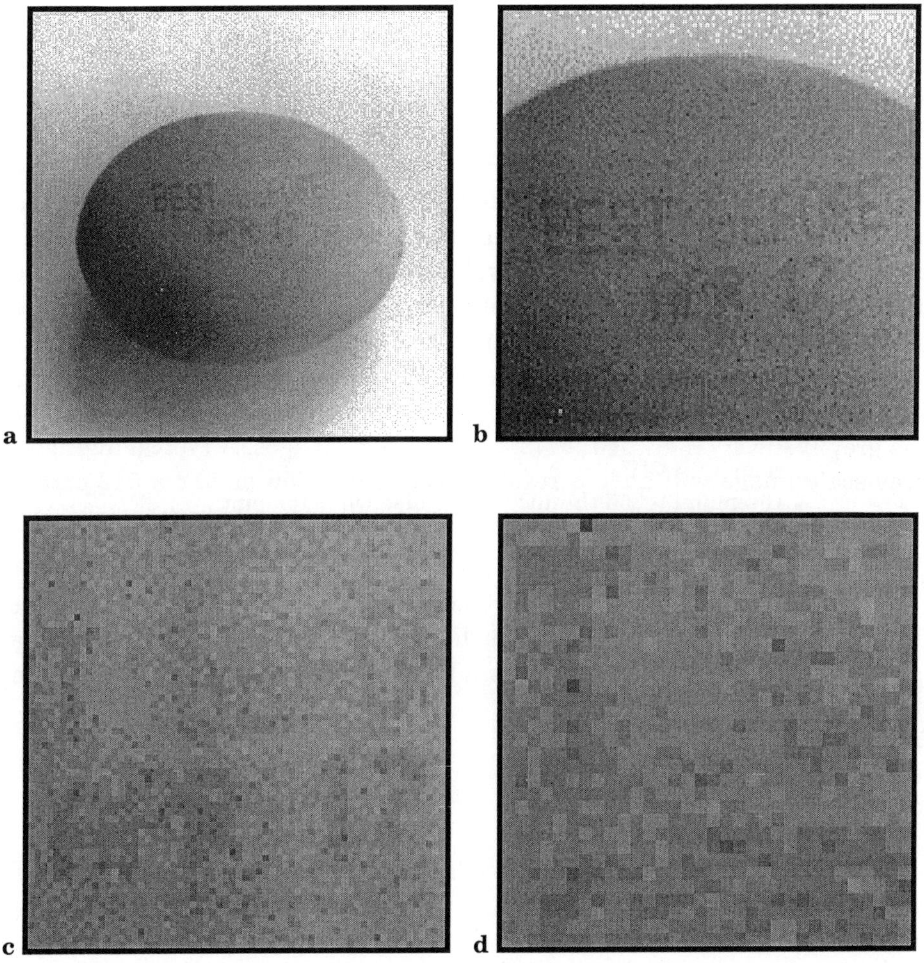

Figure 2.1 Images and zooming. (a) shows a typical 256 × 256 grey-scale image, and (b)–(d) show portions that have been enlarged by successive factors of two. Notice that as the individual pixels become larger and more readily discernible, global understanding of the image content is gradually eroded. The speckled nature of the surface of the egg and the slight highlight around the lettering also impede recognition of the characters.

It will be useful to denote this by a slightly more detailed notation, in which we are careful to state the memory spaces in which the input and output images are stored, and to refer explicitly to a general pixel in each space. For reasons which will become clear below, we will denote the intensity of the general pixel in space P by P0, and that in space Q by Q0. The above intensity doubling routine may now be expressed in the following form:

```
for each pixel in image do
   Q0 := 2*P0;
```

(Here we have employed the Pascal construct ':=' to denote 'takes the value'.)

At this point a complication in the way images containing byte variables are handled should be noted. Suppose that a pixel in the original image has a high intensity P0; then Q0 could well be greater than 255 but its value will roll around to a low value as only eight bits are available to store it. In order to avoid this we have to rewrite the above routine in the following form, making use of a temporary integer variable *tempint*:

```
for each pixel in image do
   begin
      tempint := 2*P0;
      if tempint > 255 then Q0 := 255 else Q0 := tempint;
   end;
```

A similar situation arises if we lower the intensity in such a way that it could become negative—as might happen if we subtracted 40 from the grey level values. In that case we would have to express the routine in the form:

```
for each pixel in image do
   begin
      tempint := P0 – 40;
      if tempint < 0 then Q0 := 0 else Q0 := tempint;
   end;
```

All routines which adjust the intensity of each pixel in the image according to a common rule are termed pixel operations: this category clearly includes operations such as copying images, inverting image intensities, stretching image intensities to improve contrast, and making images darker or lighter by decreasing or increasing the grey level values. It also includes the widely used operation known as thresholding, in which image regions which are lighter than a certain threshold are made completely white, and those darker than this level are made completely black (Figure 2.2); and the closely related *binarization* operation in which the binary values 0 and 1 are output. A typical routine of this type is the following:

```
for each pixel in image do
   if P0 > thr then B0 := 1 else B0 := 0;
```

Here we have taken B-space to be a binary space (which could employ bit rather than byte variables), and we have assumed that the threshold parameter *thr* is known at run time; clearly, it should have a value in the range 0 to 255. We will consider these operations again in a later section, as they are important for helping to locate and demarcate objects in grey-scale images.

2.3.2 *Window operations*

In contrast with the pixel operations described above are the so-called *window operations*: these are rather more complex than pixel operations, but are considerably more powerful. Their operation takes the following general form which makes use of a window in the input image centred about the current input pixel:

Figure 2.2 Thresholding. (a) shows the 256 × 256 grey-scale image of a sprig of parsley, and (b) shows the result of thresholding. See Figure 2.5 for a further example of thresholding.

```
for each pixel in image do
  begin
    obtain intensity values of pixels in current window;
    use these values to compute a new intensity value;
    insert new intensity value in output image;
  end;
```

The following diagram shows a commonly used labelling scheme for pixel intensities within a window around the current pixel position in image space P:

P4	P3	P2
P5	P0	P1
P6	P7	P8

This diagram makes it clear why the intensity of the current pixel was given the label P0 earlier. The following routine for removing noise from a binary image in space A illustrates how this labelling scheme is used:

```
for each pixel in image do
  if A1 + A2 + A3 + A4 + A5 + A6 + A7 + A8 = 8 then B0 := 1
  else B0 := A0;
```

If the 1's in the binary image space represent dark objects in the original grey-scale image, this noise removal operation eliminates the effect of isolated light pixels on these objects, i.e. the removal of 'salt' noise. Usually it is useful to remove both salt and pepper noise—the latter corresponding to the isolated dark pixels—and then the routine generalizes to:

```
for each pixel in image do
  if A1 + A2 + A3 + A4 + A5 + A6 + A7 + A8 = 8 then B0 := 1
  else if A1 + A2 + A3 + A4 + A5 + A6 + A7 + A8 = 0 then B0 := 0
  else B0 := A0;
```

(Clearly, some saving in computation could be made by computing the sum just once in this routine.)

We next consider how the shapes of objects can be adjusted by window operations. We can, for example, extend them to the right by applying an operation of the form:

```
for each pixel in image do
  if (A0 = 1) or (A5 = 1) then B0 := 1 else B0 := 0;
```

Similarly, we can extend objects by a single pixel in all directions using the following routine:

```
for each pixel in image do
    if (A0 = 1) or (A1 = 1) or (A2 = 1) or (A3 = 1) or (A4 = 1)
        or (A5 = 1) or (A6 = 1) or (A7 = 1) or (A8 = 1)
    then B0 := 1 else B0 := 0;
```

though the operation could be written more succinctly in the form:

```
for each pixel in image do
    if A1 + A2 + A3 + A4 + A5 + A6 + A7 + A8 > 0
    then B0 := 1 else B0 := A0;
```

In fact, this latter notation is in some ways more revealing, as it can be interpreted in the following way:

```
for each pixel in image do
    if next to an object pixel
    then mark current pixel as an object pixel
    else leave pixel value unchanged;
```

The concept 'next to' is an important one in image processing, and will appear a number of times in this monograph.

The last two operations involved extending objects in one or more directions: such operations are called dilations and contrast with erosions in which objects are contracted in various directions (see Figure 3.4(a–c)). We shall discuss these operations in more detail in Chapter 3. Meanwhile we note that window operations are able to shift object boundaries not only in binary images but also in grey-scale images. However, we illustrate a rather simpler operation on grey-scale images—that of local smoothing. Such operations are frequently carried out in order to average out noise and remove unnecessary detail from images:

```
for each pixel in image do
    Q0 := (P0 + P1 + P2 + P3 + P4 + P5 + P6 + P7 + P8)/9;
```

This is an example of a linear operation, and contrasts strongly with the non-linear operations described earlier: in fact, non-linear operations on binary images are probably best described as logical operations (even when they are performed with the aid of additions, they still involve logical *if* statements, and hence count as non-linear operations). We can go further and state that the above linear operation is in fact a convolution, a class of operation which is both linear and uniform over the whole of the data space. We do not dwell on this here, except to write down a general operation of this form, each pixel intensity within the window being assigned its own individual weighting factor:

```
for each pixel in image do
    Q0 := c0*P0 + c1*P1 + c2*P2 + c3*P3 + c4*P4
            + c5*P5 + c6*P6 + c7*P7 + c8*P8;
```

Although this is an unambiguous statement of the operation, it is slightly clumsily expressed, and is hence usually formulated in the form of a convolution mask:

$$\begin{bmatrix} c4 & c3 & c2 \\ c5 & c0 & c1 \\ c6 & c7 & c8 \end{bmatrix}$$

In this form the linear smoothing operation can be expressed in terms of the convolution mask:

$$\frac{1}{9}\begin{bmatrix} 1 & 1 & 1 \\ 1 & 1 & 1 \\ 1 & 1 & 1 \end{bmatrix}$$

Convolutions represent an extremely powerful way of processing images, and convolution masks have been devised for a wide variety of tasks, including (a) simple smoothing (as above), (b) Gaussian smoothing (with mask weights which more closely match a 2D Gaussian profile), (c) edge detection, (d) line segment detection, (e) hole and spot detection, (f) corner detection, and many more. However, the fact remains that they are linear operations, and hence there is a whole range of tasks which they cannot do. For example, as we shall see below, convolutions designed for smoothing images are unable to cope well with impulse noise, and have the added disadvantage that they blur images as well as eliminating noise. Another example is provided by binary images, there being no pure binary convolution operations (combining 0's and 1's arithmetically is bound to lead to fractional values or else values outside the range 0 to 1). This means that non-linear operations will need to be core elements of our image processing methodology.

2.3.3 *Some problems and implementation issues*

2.3.3.1 *Eliminating underflow and overflow*

We noted earlier that computations of grey levels can result in intensity values outside the range 0 to 255, and that steps may be required to ensure that the results are kept within these bounds. For rigour and safety (in view of amendments that may be made to algorithms to help them cope with real data), it is advisable to include a unified conditional statement in all routines, at the final stage before leaving the window operation:

```
for each pixel in image do
   begin
      compute tempint;
      if tempint > 255 then Q0 := 255
      else if tempint < 0 then Q0 := 0
      else Q0 := tempint;
   end;
```

2.3.3.2 *Solving image border problems*

When window operations are applied to whole images, the problem arises of what values to assume for window elements which are outside the image. In general, there is no complete solution to this problem, as there is no information about what is happening outside the image. Broadly, there are two ways of tackling the problem. First, we can make the 'safe' assumption that there are no objects just outside the image: for a binary image this amounts to taking all values outside the image as 0; then performing a dilation operation will not introduce into the image any objects for which there is no available evidence, and errors will be kept at a minimum. Similarly, for an erosion operation, we can take all pixel values outside the image to be 1, again keeping errors at a minimum for objects which are partly obscured at the boundaries of the image. For grey-scale images a strategy of assuming that all values outside the image are 0 or 255 would introduce enormous errors, but a useful strategy is to take the value as being equal to that of the *nearest* pixel within the image. All such strategies are prone to error, but the basic idea has to be to reduce the resulting error to a minimum, and particularly not to add artefacts for which there is no evidence within the image.

The strategies outlined above do, however, contain some risk. To avoid risk, intensity values for which there is incomplete evidence should not be inserted into the output image. This implies that image border pixels should be left blank, and that the basic window statement should be changed from "**for** each pixel in image **do**" to "**for** each non-border pixel

do". However, when monitoring image processing operations, it is a good idea to make the border pixels black rather than blank, so that, when looking at the processed image, it is obvious which parts of the data can be relied upon.

While this strategy seems safer, it can give rise to problems if a sequence of window operations is applied in sequence—as often happens in practical applications. For example, we might wish to remove noise, then threshold the image, then erode the objects a number of times, then dilate again to modify the shapes, then perform a skeletonization operation (see Chapter 3). Unfortunately, this means that the valid region of the output image can become smaller and smaller, or even (theoretically) non-existent! At the very least it can mean that an inspection algorithm will miss a substantial, relevant portion of the image. Clearly, if N operations of this type occur, the blank border region will be N pixels wide. However, this assumes that 3×3 windows are used for image processing: in fact, windows are often considerably larger than this—sometimes (notably in texture analysis tasks) up to 17×17. So the active region of the image may be substantially depleted. Fortunately, some algorithm design strategies are able to limit these problems: one example is that Hough transform approaches can infer the presence of objects, even when they are partially occluded at the boundaries of the image; we shall investigate this aspect more fully in Chapter 4.

2.3.3.3 Sequential and parallel processing

All the routines presented so far have involved *parallel processing*, i.e. they could well have been performed in parallel at each pixel location, since the processed image has always been separate from the input image. Indeed, if this had not been the case, it would in general have mattered in which order the pixels were processed, and different types of scan—such as forward raster scan and reverse raster scan—would have given different results. In fact, there are some classes of processing which escape from these problems. First, pixel operations are evidently immune from them, but not general window operations. Second, amongst the window operations, binary salt and pepper noise removal is immune, but erosion operations are not. In general, many of the basic image processing concepts such as erosion, dilation, edge detection, and grey-scale noise elimination are parallel in nature, and care has to be taken to ensure that the output image is inserted into a different image space. Nevertheless, there are occasions when it is useful to employ *sequential processing* (in which new values are immediately re-inserted into the original image space)—as we shall see in Chapter 3: in such cases the algorithms need to be quite carefully designed in order to avoid undesired effects such as anisotropy.

2.3.3.4 *The need for restoring operations*

The fact that many window operations involve parallel processing is inconvenient, since a sequence of standard routines, all of which have the same input space and the same destination space, will not work correctly together, and will have to be rewritten with new input and destination spaces. An alternative is to restore all processed images to the same input space using pixel copying operations, though these will incur additional computational overheads. We by-pass this problem here, noting that in principle it can be solved with the help of an intelligent optimizing compiler, which would merely relabel all the spaces appropriately.

2.3.3.5 *Speed of processing*

It has long been known that applying a sequence of basic operations is potentially less computation intensive than applying a single more complex operation. Thus applying a sequence of four 3×3 window operations can be more economical than applying a single ideal 9×9 window operation, e.g. for eroding all the objects in an image by four pixels. Broadly, the saving in computation is roughly in proportion to the sums of the areas of the windows that are involved, viz. 81:36 in the above example. Sometimes higher savings can be achieved, as when a 2D Gaussian smoothing operation is implemented as two 1D smoothing operations which act along the respective image axes: in that case the 9×9 mask is replaced by a 1×9 and a 9×1 mask, and the saving is roughly 81:18. Space does not permit this aspect to be investigated further here, but savings in computation have to be considered very carefully in view of the large numbers of pixels that have to be processed per second in most real-time applications.

2.3.3.6 *Processing in the spatial frequency domain*

In many applications it is attractive to transform data into a different form so that it can be processed more conveniently. We have already mentioned the Hough transform in this context. The Fourier transform is very well known in many scientific contexts—not least in signal processing. In particular, if we wish to remove temporal noise by low-pass filtering, we must first convert the data to the frequency domain, perform the filtering, and then convert back to the time domain. Similarly, if we wish to remove spatial noise from images, we must first convert the image data to the spatial frequency domain, then perform the low-pass spatial frequency filtering operation, and finally convert back to the spatial domain. Using this approach is efficient when very low spatial frequencies are involved. However, when the cut-off frequencies are high, it is more efficient to use small window convolution masks to perform the filtering. This fact limits the value of Fourier transformations in the sort of applications envisaged in this monograph.

2.4 Median and rank-order filters

It was stated in the previous section that although linear smoothing algorithms may be used to remove noise from images, they do not cope well with impulse noise and have the added disadvantage that they introduce blur. These problems do not arise with the median filter, which works by eliminating extreme intensity values, and indeed selects the middle intensity value from amongst those in each window. The basic algorithm is shown in Table 2.1.

It will be seen that the algorithm starts by clearing a histogram, and then incrementing those storage locations for which intensity values are found within the window (incrementation is indicated by the '++' symbol). It then progresses through the histogram until it has found the central intensity value, and returns this value as the median. The algorithm is quite slow because of the need to clear a histogram of 256 locations, and because of the need to progress through about half of them to locate the median. There are many ways of speeding up the algorithm, both in software and in hardware, but we do not need to consider them here. Suffice it to say that the median filter represents the single most important algorithm for removing noise—particularly impulse noise—in grey-scale images, mainly because it is simple and causes negligible blur. It has, however, been shown to cause detectable shifts of curved object boundaries, and means have been sought of eliminating this problem. Such shifts are potentially serious for inspection and measurement: in such situations, either median filtering should be avoided, or an appropriate correction

Table 2.1 Basic median filter algorithm

```
for each non-border pixel do
   begin
      for i := 0 to 255 do hist[i] := 0;
      hist[P0] ++; hist[P1] ++; hist[P2] ++; hist[P3] ++; hist[P4] ++;
      hist[P5] ++; hist[P6] ++; hist[P7] ++; hist[P8] ++;
      i := 0; sum := 0;
      while sum < 5 do
         begin
            sum := sum + hist[i];
            i ++;
         end;
      Q0 := i – 1;
   end;
```

made to each measurement (Davies 1989a, 1997e).

Rank-order filters are similar in operation to median filters, and indeed the median filter is a particular type of rank-order filter in which the median (middle) intensity is selected rather than some other value in the ordered set of integers. In fact, there are nine possible rank-order filters for a 3×3 window: any of these may be formed by replacing the five in the **while** statement by another integer in the range one to nine. If the five is replaced by a one, the filter selects the lowest intensity value, and gives rise to a *min filter*: if it is replaced by a nine, the filter selects the highest intensity value, and gives rise to a *max filter*. Figure 2.3 gives examples of the operation of all three types of filter: however, the filters used here employ a window size of 7×7, so in this case they select the first, twenty-fifth and forty-ninth of the ordered intensity values.

To understand the properties of rank-order filters, we must first note that the most extreme intensity values are those most likely to be due to noise. Hence min and max filters are liable to give rise to noisy or distorted images (in fact, a min filter will tend to produce dark patches around locations where there were dark noise pixels in the original image, while a max filter will tend to produce light patches around locations where there were light noise pixels). However, rank-order filters whose order values are closest to the median will be relatively good at eliminating noise. Clearly, rank-order filters other than the median will tend to pull the image either downwards or upwards in intensity (i.e. make the image darker or lighter). An interesting point is that min and max filters are the grey-scale counterparts of binary erosion and dilation operators[1], which can themselves be reformulated as min or max operations on binary window values.

2.5 Thresholding

As remarked in Section 2.3.1, thresholding is widely used to help with the location and demarcation of objects in grey-scale images. The basic idea is to view objects against a contrasting background so that they have different sets of intensity values in the input image; in that case it should be possible to select an intensity threshold which will permit separation to be achieved.

Ideally, the threshold should be selected automatically by computer. One means of achieving this involves constructing an intensity histogram, as in the following algorithm:

```
for i := 0 to 255 do hist[i] := 0;
for each pixel in image do
    hist[P0] ++;
```

[1] However, for grey-scale images rather more general dilation and erosion operators exist —see Chapter 3.

Figure 2.3 Rank-order filtering. (a)–(c) show the respective effects of median, min and max filtering on the egg image of Figure 2.1. Notice how the median (a) removes noise (in this case surface speckle), while the other two filters exaggerate it in various ways: the lettering is similarly affected. In contrast with the basic algorithm of Table 2.1, all three filters employed here operate within a 7×7 window.

Next, some means of analysing the histogram has to be devised. Ideally, if objects have been made to appear dark against a light background, there will be two peaks in the histogram, the one at low intensity corresponding to the objects, and the other at high intensity corresponding to the background. However, variations in illumination and in the surface brightness of the objects—and also image noise—will prevent the histogram profile from being smooth, and there will be quite a number of local minima and maxima (Figure 2.4). To overcome these problems it is possible to smooth the histogram until only two peaks appear: then the algorithm only has to find the first maximum followed by the first minimum on progressing from one end of the histogram:

```
i := 0;
repeat
   i ++;
until (hist[i + 1] < hist[i]) or (i >= 255);
repeat
   i ++;
until (hist[i + 1] > hist[i]) or (i >= 255);
if i >= 255 then writeln('no proper minimum');
```

This approach is beautifully simple, but is unlikely to work well with most real images as the amount of noise on the unsmoothed histogram is liable to be excessive. Instead, procedures which search for two main peak positions and then search for a global minimum between them, finally refining its position by noise removal operations, are likely to be more reliable. Whatever approach is adopted, the basic principle is that of finding a *significant* minimum in the histogram.

Unfortunately, a number of other problems can arise which make threshold selection difficult. Perhaps the main problem is the fact that most images contain a variety of objects and background clutter, while the objects themselves contain a variety of relevant features. This means that the histogram will be intrinsically multimodal, and the wrong threshold value may well be computed. It will be clear that any algorithm that obeys simple rules for analysing the histogram rather than aiming to understand what is in the image is to some extent *ad hoc*, and will be liable to arrive at erroneous thresholds: in some cases thresholds could be considerably in

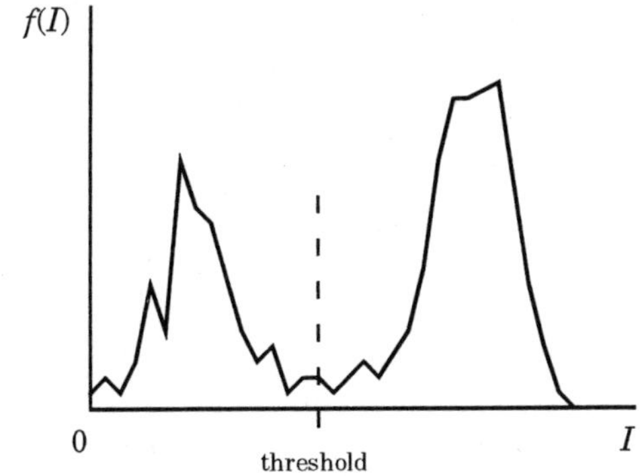

Figure 2.4 Finding a threshold. This diagram shows a somewhat noisy intensity histogram, making it difficult to find an optimum threshold position.

error, as might happen when more objects of a certain type appear in the image than envisaged when the histogram analysis procedure was devised, so that a distinct reduction in the amount of normal background is evident. However, in a restricted (e.g. production line) environment, additional rules which constrain the final permitted threshold values might solve the problem.

A further problem which should be mentioned is that variations in the sizes of the main peaks in the histogram will shift the minimum position one way or the other and hence bias the choice of threshold level. Much work has been done on this effect and it can be allowed for. One approach is to construct the intensity histogram by looking only at pixels which are near the edges of objects, though this involves prior application of an edge detector (see below) and makes the whole process more complex.

Finally, variations in illumination across an image can sometimes be so great that they not only broaden the main peaks in the histogram, but also merge them together so that no minimum is detectable: in other cases the minimum is detectable but cannot accurately be determined. However, the problem is deeper than this, as such a minimum is effectively meaningless: a different threshold is really needed for each location in the image. To solve this problem, adaptive thresholding is required: this will be discussed in the following section.

2.6 Adaptive thresholding

Perhaps the most obvious method for solving the problem of non-uniform illumination is to model the background in the absence of any objects, and to subtract the resulting intensity pattern from that observed when objects are present. In principle, the difference image can then be analysed and thresholded as outlined above. While this approach can be made to work reasonably well in constrained situations, for example with known flat objects such as biscuits on a conveyor, in more complex situations it tends to be plagued by problems such as the effects of shadows and secondary lighting (i.e. light reflected from one object to another). Thus a more rigorous approach which determines the thresholding levels from the current image is required.

As the fundamental requirement is to select thresholds which are suitable for the individual local regions, a reasonable strategy should be to divide the image into square regions and construct an intensity histogram for each region, and then proceed as before. There are several problems with this approach: (1) the results obtained from different regions will not meld together properly, and steps will appear in the boundaries of objects wherever a change in threshold has occurred at the boundary of a region; (2) if the regions are too small the statistics will be poor and the thresholds will be determined inaccurately, whereas if the regions are too large, the thresholds will not be optimal throughout the regions; (3) in many cases there will be no objects in some regions and no background in others, a situation which will lead to difficulty and in some cases totally erroneous thresholds being produced. These problems have been tackled by using suitably chosen region sizes, and introducing interpolation schemes which check or compute valid thresholds for all regions, and also local corrections to them for pixels not at region centres (Chow and Kaneko 1972). Such schemes are not perfect: in addition, they tend to be computation intensive. However, there are applications where a simpler type of solution can be employed—namely the local thresholding approach to be described next.

What is really wanted to solve this problem is a local statistic—i.e. some easily calculable quantity—which permits an acceptable threshold value to be computed fairly rapidly and in a manner which will not vary too much from one pixel to the next: if this can be achieved, undesirable steps should not be produced on the boundaries of objects.

A simple but useful statistic is the mean of all the pixel intensities within a 3×3 or slightly larger window. This will tend to fall between the background and object intensity values, but only where the window includes pixels which lie within both the background and the foreground objects. In such cases, the mean of the maximum and minimum intensities within the window will provide an even better statistic, as it will not be sensitive to the precise numbers of background and foreground pixels within the window. However, such a statistic will be sensitive to impulse noise, and care should be taken in its use. This type of statistic is useful when there is a reasonable balance between foreground and background pixels: this does not apply when objects are large or few and far between, but is an excellent approach when there are lots of small objects; it could for example be used to help with sorting nails, or for thresholding a printed page prior to applying OCR algorithms, or for fingerprint analysis (Figure 2.5). Clearly, such statistics can only be used with some algorithmic device which notices when there are no background or foreground pixels within the window. However, it is relatively straightforward to construct such a device, either by placing absolute limits on the final threshold, or by limiting how much the threshold can change from one pixel to the next, or else by ensuring that there is sufficient difference between the maximum and minimum intensities (as in Figure 2.5(b)). In the first of these cases, the final algorithm will have the basic form (Davies 1997e):

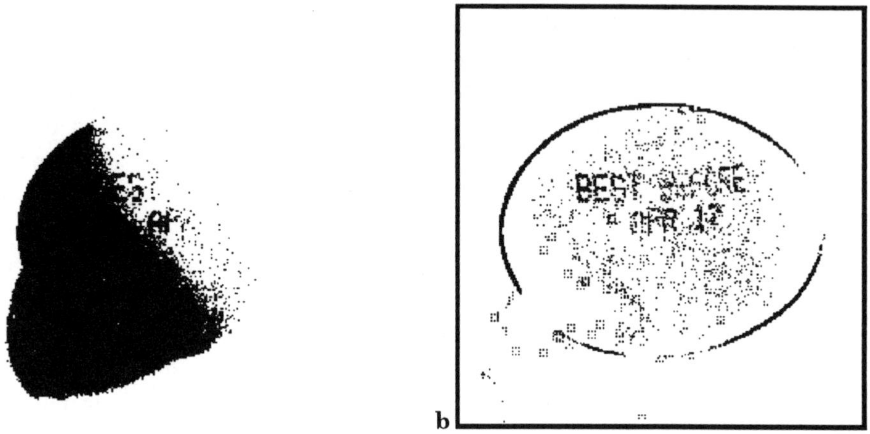

a b

Figure 2.5 Adaptive thresholding. (a) shows the result of applying a global threshold to the egg image of Figure 2.1, and (b) shows the effect of applying an adaptive threshold based on the use of the min + max statistic. Note the somewhat surprising improvement resulting from the use of this simple algorithm: any lack of perfection is due more to the limitations of the data than to the algorithm itself (see Figure 2.1).

```
for each non-border pixel do
   begin
      find min and max intensities in window;
      if min > 100 then thr := 100
      else if max < 100 then thr := 100
      else thr := (min + max)/2;
      if P0 < thr then B0 := 1 else B0 := 0;
   end;
```

A variant on this approach is that of finding the mean and standard deviation (or more conveniently, the mean deviation) in a window, and computing the minimum or maximum likely intensity value; then intensities outside the likely intensity range would constitute objects or defects which should be thresholded for further analysis. This method can be used for applications as different as locating stars in astronomical images, defects on sheets of paper, or dark cracks on eggshells. For this last case, the final algorithm will be of the basic form[2]:

[2] For bright objects such as stars, the sign accompanying the mean deviation would have to be changed in the expression giving the threshold value.

```
for each non-border pixel do
  begin
    find mean and meandev intensity in window;
    thr := mean – 2*meandev;
    if P0 < thr then B0 := 1 else B0 := 0;
  end;
```

In the case of problems arising from noise, an image can be formed using the thresholds computed by these simple algorithms, and a smoothed form of this image used to provide an improved set of threshold values. Such approaches can be made to work well in quite a range of special cases. However, they run into problems with complex images, including those where the objects have features with a variety of intensity values.

2.7 Edge detection

While thresholding is a highly efficient way of segmenting objects, features and defects from the background, it only works well in a restricted range of applications, e.g. when the lighting is carefully set up and the objects and background are restricted in the range of intensity values they produce. Clearly, adaptive thresholding schemes widen the range of applicability of this approach, but there are limits beyond which parameters other than intensity must be called into play. Intensity gradient is one such parameter. However, if this parameter is to be computed at every pixel location, why not use it as a segmentation parameter in its own right, rather than trying to use it with the basic intensity parameter? If this were possible, it could result in an attractive new segmentation paradigm rather than one which is more complex and unwieldy. This section considers how this possibility can be realized.

The strategy for locating objects via their edges is necessarily different from that when thresholding is used. In the latter case the objects are made to stand out as a whole: this may be described as a region-based approach to segmentation. Locating objects by their edges focuses on object boundaries rather than object regions. Hence we should aim to devise means of highlighting edges and actually ignoring regions. This can be achieved by searching for pixels which lie at positions where intensity gradient is much larger than normal.

A possible approach is to apply convolution masks (Section 2.3.2) which enhance edges of various orientations. For example, we could enhance vertical edges by applying a mask of the form:

$$\begin{bmatrix} -1 & 0 & 1 \\ -1 & 0 & 1 \\ -1 & 0 & 1 \end{bmatrix}$$

Similarly, horizontal edges could be enhanced by a mask of the form:

$$\begin{bmatrix} 1 & 1 & 1 \\ 0 & 0 & 0 \\ -1 & -1 & -1 \end{bmatrix}$$

and diagonal edges could be enhanced by masks such as:

$$\begin{bmatrix} 0 & 1 & 1 \\ -1 & 0 & 1 \\ -1 & -1 & 0 \end{bmatrix}$$

Note that four of the eight masks of these types differ from the others only in polarity: hence edges of all orientations can be located using just four masks. However, it should be noted that edges which are not exactly at 0°, 45°, 90°, 135°, ... to the image positive x-axis will give slightly reduced responses.

It has long been known that the above template matching strategy is inefficient in requiring more masks than absolutely necessary: it is also responsible for the reduced response away from the 0°, 45°, 90° or 135° orientations; and finding which mask gives the largest response is only capable of determining orientation within ± 22.5°. A better strategy involves noting that intensity gradient is a vector **g** with magnitude g and direction θ, and is fully determined by *two* orthogonal components (Haralick 1980). Thus we can use the first two masks above to estimate the x and y components of intensity gradient, g_x and g_y. We can then calculate the intensity gradient and orientation using the respective formulae:

$$g = (g_x^2 + g_y^2)^{1/2} \tag{2.1}$$

$$\theta = \arctan\left(g_y / g_x\right) \tag{2.2}$$

Unfortunately, this last formula only permits orientation to be estimated to an accuracy of about 5°. As we shall see in Chapter 5, this is insufficient for some purposes. However, the situation can be rectified by using the Sobel masks:

$$\mathrm{S}_x = \begin{bmatrix} -1 & 0 & 1 \\ -2 & 0 & 2 \\ -1 & 0 & 1 \end{bmatrix} \qquad \mathrm{S}_y = \begin{bmatrix} 1 & 2 & 1 \\ 0 & 0 & 0 \\ -1 & -2 & -1 \end{bmatrix}$$

The fact that the coefficients in these square masks emulate a circular shape permits them to act more isotropically and to estimate orientation to better than 1° (Davies 1984b, 1997e): in fact, this degree of accuracy is quite remarkable considering how little intensity information is available within the 3×3 windows.

We next consider how to use the information provided by the magnitude g of the intensity gradient (Figure 2.6). If an image is constructed with pixel values equal to the local values of g, this may be thresholded at a convenient level, and can immediately provide a binary image containing the edges of all the objects in the image (Figure 2.7). This process tends to be less critical than thresholding the intensity to highlight the object regions, since it automatically ignores slow variations in the illumination of the objects in the scene. Nevertheless, there is a tendency for edges to peter out in some places and to be rather broad in others. To some extent this can be corrected by altering the thresholding level, and in many situations this is an adequate solution. However, lowering the threshold until all the edges are included will tend to broaden high contrast edges, and sometimes so much accuracy is lost that this is not a viable approach. Fortunately, applying Hough transform and related model based approaches to object location can largely overcome these problems (see Chapter 4). When applying such methods it is often better to optimize accuracy by using relatively high threshold levels even though this can cause the fainter edges to be eliminated.

Figure 2.6 Edge enhancement. This figure shows the result of applying the Sobel edge enhancement operator to the sprig of parsley of Figure 2.2.

2.8 Concluding remarks

This chapter has come a long way. After discussing basic properties of images it showed how image processing could be achieved by pixel operations and by window operations; it also dealt with a variety of implementation problems. It emphasized that image processing operations can be linear (convolutions) or non-linear, both types of processing having their advantages and limitations, though it would be natural to use various

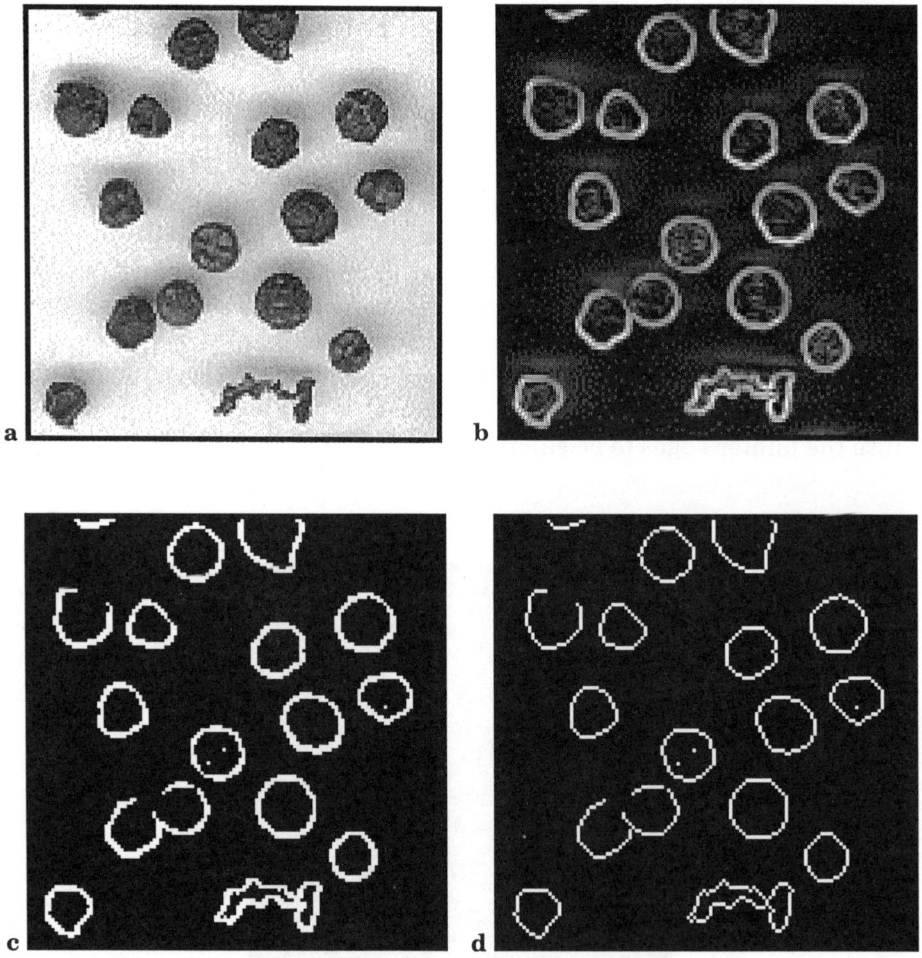

Figure 2.7 Edge detection applied to a peppercorn image. (a) shows the original 128 × 128 grey-scale image; (b) shows the result of edge enhancement using a Sobel operator; (c) shows the result of thresholding to perform edge detection; and (d) shows the result of thinning the detected edges. Notice that, as a result of low contrast in certain locations, the edges in (c) and (d) are not always fully connected around the object boundaries.

combinations of these operations in order to achieve specific goals in most practical situations.

Noise elimination is an important area of image processing, though it might be thought that it would have limited value in applications such as automated visual inspection. However, it turns out that one method for achieving it—namely median filters—can be used in the detection of corners, and these are important in the location of certain types of object. In addition, rank-order filters (which generalize the median filter concept) are able to achieve a number of shape modification functions, a property they have in common with so-called morphological functions such as erosion and dilation. It has been seen that all these functions are easily implemented as window operations. At this point we have moved on to shape analysis, a topic which is studied in more detail in the following chapter. In particular, we shall see that shape analysis is of especial value for recognition.

However, it is logical to locate objects before trying to recognize them. This function is achieved by image segmentation. Early in this chapter we found that in ideal situations objects can be segmented from the background by simple thresholding operations, though further study indicated that this process is often far from straightforward: not only is automatic threshold determination difficult when the intensity histogram contains a lot of noise, but also it may exhibit a number of independent peaks and troughs, which may merge together without warning from image to image; this means that threshold determination algorithms, which are *ad hoc* in the sense that they do not 'understand' the content of the image in any meaningful way, are prone to arrive at erroneous conclusions. Equally important, when the scene being observed is prone to variations in illumination, it is meaningless to allocate any single threshold value over the whole image, and so adaptive thresholding procedures are required. In certain cases these can operate extremely effectively, but when this is not so, additional image parameters such as intensity gradient must be called into play to help with segmentation. However, on the basis of not making any method more complicated than necessary, an alternative is to employ edge detection operators which start by computing the intensity gradient. In general this approach provides a far more widely applicable solution than thresholding to the problem of segmentation.

At this stage we have moved from a general perusal of image processing to a study of the intricacies of image segmentation. It is worth pointing out that the starting assumption of region-based segmentation—that objects may be segmented by searching for regions of uniformity in images—is deeply flawed: there is no *a priori* reason why uniformity should provide a rigorous basis for segmentation, and this is the reason that the thresholding approach so often fails. Similarly, there is no reason why the boundaries of real objects should be straightforwardly located by edge detection algorithms, which merely look for high-contrast boundaries between regions. On the contrary, this approach too is prone to failure, for it is observable that in real scenes, parts of objects will merge with their backgrounds, reducing edge contrast to zero and preventing edge detection algorithms from producing continuous boundaries around objects. Thus, whichever route to segmentation is adopted, it is necessary to have means for inferring the presence of objects from incomplete information provided by the low-level methods described in this chapter. How this is achieved will be discussed in some detail in Chapter 4.

As anticipated in the introduction, this chapter has been unable to cover the whole subject exhaustively. However, its storyline has dealt with a variety of interlocking concepts and has shown something of the nature of image processing. It will be up to the remaining chapters in Part 1, and also more advanced texts, to cover some of the issues in more depth. In particular, Chapter 3 deals with shape analysis, including skeleton analysis and morphological operators; Chapter 4 covers feature detection and model-based shape recognition using techniques such as Hough transforms and graph matching; while Chapter 11 includes a detailed discussion on threshold selection as applied to X-ray images. Finally, Davies (1997e) goes into rather greater depth on a variety of relevant issues, including thresholding techniques, edge detection, boundary pattern matching, and the rationale and application of Hough transforms.

Chapter 3
Shape analysis

3.1 Introduction

Shape analysis is perhaps the most practically useful area of image processing, and is very important in automated visual inspection and other aspects of food processing. This is so whether 2D or 3D shapes are involved, but in this chapter we shall concentrate on 2D shape analysis. We have seen in the previous chapter how grey-scale images can be thresholded to give binary images: the latter contain all the information necessary for analysing the sizes and shapes of 2D objects. In what follows we class size and shape analysis under the generic heading of shape analysis, and normally the same techniques are appropriate for studying both.

In a single chapter it will be impossible to cover the whole subject of 2D shape analysis. Nevertheless, by discussing a number of widely-used techniques we shall be able to give something of the flavour of the subject, and to provide the basis for the application of image processing to food manufacture to be undertaken in Part 2. As hinted in Chapter 2, there are two main approaches to the subject—region-based analysis and boundary-based analysis. We shall start by considering the first of these.

3.2 Connected components analysis of images

It is always worth remembering that when computers analyse digital images, they do not immediately 'see' the answer in the way we do. Indeed, our capability for doing so is in many ways an impediment to devising vision algorithms. In particular, when an image is filled with objects of various types, sizes and shapes, one of the major problems is that of disentangling the data and separating the shapes, taking each one to be a connected entity. For this reason it is very important to have sound algorithms for distinguishing objects which are connected together from those which are not. In this context it has to be assumed that if adjacent pixels are both 1's, they must both belong to the same *connected component* (though it may be possible at a later stage for some algorithm to infer that a single connected component results from two or more overlapping or touching objects and to separate them).

In examining connectedness within a digital lattice of pixels, a well-known problem arises. Imagine a thin diagonal line cutting right across an image. Taking this line to be a connected component, it will necessarily divide the background into two separate unconnected regions[1]:

```
. . . . . . . . 1 . . . .
. . . . . . . 1 . . . . .
. . . . . . 1 . . . . . .
. . . . . 1 . . . . . . .
. . . . 1 . . . . . . . .
. . . 1 . . . . . . . . .
. . 1 . . . . . . . . . .
. 1 . . . . . . . . . . .
1 . . . . . . . . . . . .
. . . . . . . . . . . . .
. . . . . . . . . . . . .
. . . . . . . . . . . . .
```

Now notice that a line in the background running along an oppositely orientated diagonal will pass between the 1's on the line and constitute a connected entity in the background. Thus we have two separated background regions which are actually joined through the line that separates them. This paradox can only be resolved if the assumption about connectedness is rescinded, either in the foreground or in the background. It is actually more usual to assume that:

- diagonally adjacent 1's (in the foreground) are connected
- diagonally adjacent 0's (in the background) are not connected.

Alternatively, we can state that the foreground is '8-connected' and the background is '4-connected'.

The concept of connectedness is right at the heart of discussions on object labelling, counting, measurement and containment (the last of these pertaining to location of small objects within holes in other objects). There are two main approaches to connectedness problems. In the first, normal parallel processing procedures extract and maintain sufficient information during image scans so that they can arrive at unique labellings of all the objects, or perform the other relevant tasks mentioned above. In the second, tracking algorithms extract and analyse the shapes of all the objects in the image: we shall defer discussion of this latter approach until later in this chapter.

Labelling can be performed by the general type of procedure shown in Table 3.1. The reason for keeping a record of label adjacencies is that

---

[1] To avoid confusion, background 0's will normally be denoted by dots in this chapter.

Table 3.1 Object labelling procedure

```
clear count parameter;
clear label adjacency table;
for each non-border pixel do
    if current pixel is adjacent to one or more labelled pixels
    then
        begin
            give current pixel lowest adjacent label;
            record all label adjacencies;
        end
    else
        begin
            give current pixel next unused label;
            increment count parameter;
        end;
analyse all label adjacencies encountered in scan;
produce a lookup table giving optimum labelling;
relabel objects appropriately;
```

different parts of the same object can initiate their own labels, but at the point that they become adjacent within the object these different labels must be recorded as arising from the same object (Figure 3.1). Resolving these separate labels can be achieved by abstract processing of the data in the label adjacency table, when the latter is complete, or else by repeated scanning over the image, sweeping lower and lower labels over the objects, until no further change takes place. This latter procedure can take significant computational effort, though this can often be alleviated by employing different types of scan—forward raster scan, reverse raster scan, reflected forward raster scan, and so on. However, excessive image noise tends to make the problem worse than might be expected, and it is better to carry out abstract processing of the data in the label adjacency table than to perform repeated image scans.

It should be noticed that, as the object labelling procedure of Table 3.1 maintains a count parameter, it automatically counts the objects in the image as well as labelling them. On the other hand the procedure ignores containment, merely counting and labelling contained objects as any other objects. Thus containment will have to be determined by querying each object in turn to determine whether its holes contain other objects, or by some other strategy such as finding which objects are accessible by propagating through the background from the boundaries of the image. In the present context, *propagation* is defined as a process which passes some property from pixel to pixel via a valid adjacency path—8-connected for objects, 4-connected for background; clearly, the object labelling procedure of Table 3.1 involves propagation.

```
· · · · · · · · · · · · · · · · · · · · · · · · · · · · · ·
· · · · · · · · · 1 · 2 · · · · · · · · · · · 3 · · · · · ·
· · · 4 4 4 · · · 1 1 1 · · · · · 5 5 · · · 3 3 3 · · · · ·
· · 4 4 4 4 4 · · · 1 · · · · · 5 5 · · · · · 3 · · · · · ·
· · 4 4 4 4 4 · · · · · · · · · · 5 5 · · · · · · · · · · ·
· · 4 4 4 4 4 · · · · · · · · · · · · · · · · · · · · · · ·
· · · 4 4 4 · · · · · · · · · · · · · · · · · · · · · · · ·
· · · · · · · · · 6 6 6 6 · · · · · · · · · · · · · · · · ·
· · · · · · · 6 6 6 6 6 6 6 6 · · · · · · · · · · · · · · ·
· · · · · · · 6 6 6 6 6 6 6 6 6 · · · · · · · · · · · · · ·
· · · · · · · · 6 6 6 6 6 6 6 6 6 · · · · · · · · · · · · ·
· · · · · · · · · 6 6 6 6 6 6 6 6 · · · · · · · · · · · · ·
· · · · · · · · · · 6 6 6 6 6 6 6 6 · · · · · · · · · · · ·
· · · · · · · · · · · 6 6 6 6 6 6 6 · · · · · · · · · · · ·
· · · · 7 7 7 7 7 7 6 6 6 6 6 6 6 6 · · · · · · · · 8 8 · ·
· · · · 7 7 7 7 7 6 6 6 6 6 6 6 6 6 · · · · · · · 8 8 8 · ·
· · · · · 7 7 7 6 6 6 6 6 6 6 6 6 · · · · · · · 8 8 8 · · ·
· · · · · 7 7 6 6 6 6 6 6 6 6 6 6 · · · · · · · 8 8 · · · ·
· · · · · · 6 6 6 6 6 6 6 6 6 6 · · · · · · · · · · · · · ·
· · · · · · · 6 6 6 6 6 6 6 6 · · · · · · · · · · · · · · ·
· · · · · · · · · 6 6 6 6 · · · · · · · · 9 9 9 · · · · · ·
· · · · · · · · · · · · · · · · · · · · · 9 9 9 · · · · · ·
· · · · · · · · · · · · · · · · · · · · · · · · · · · · · ·
```

Figure 3.1 Object labelling. This diagram shows how a simple labelling algorithm copes with a scene containing an assortment of cakes. Note how two objects have label clashes which could be corrected by a single further reverse scan. With more complicated shapes, further scans might be needed.

3.3 Skeletons and thinning

The shapes of connected components can be analysed in a variety of ways, but early on in the development of the subject the concept of skeletonization assumed great importance. *Skeletonization* is the process of creating the skeleton of a shape, whereby objects with limbs of nearly constant width are represented instead by medial lines (Figure 3.2). Thus the often irrelevant limb widths are ignored, and limb lengths, positions and junctions are recorded instead of more detailed information about the boundary of the object. In fact, object boundaries and object skeletons form alternative but ultimately equivalent shape representations, though the skeleton is attractive precisely because so much redundant information has been eliminated. Nevertheless, it retains all the connectedness information of the original object, and can thus be used to ascertain numbers of limbs, holes and other parameters of interest. On the other hand, guaranteeing maintaining all the connectedness information is a stringent requirement which some algorithm developers have found excessively difficult! It should also be remarked that skeletonization algorithms are quite intricate and there is a need for rigour in checking that typographical errors have not been introduced: this situation contrasts with that for many grey-scale algorithms (such as edge detectors) for which there is normally significant redundancy which will prevent such errors from leading to catastrophic failures.

Historically, the most important skeletonization procedure has been thinning, which involves stripping off layer after layer of edge pixels until only a connected unit width stick figure (the skeleton) remains. In Section 2.3.3.3 we noted that sequential processing of images can be valuable in certain cases where there is significant dependence on order of processing. This applies particularly to thinning, as simultaneous removal of more than one edge pixel can disconnect limbs, so in principle every removal of a pixel needs to be checked independently for potential disconnection. However, such a step by step approach to thinning can be inefficient. Perhaps more importantly, it can result in a biased skeleton—at least if a simple forward raster scan is adhered to. For this reason, parallel algorithms are most widely used for thinning, though they tend to retain a vestige of sequential processing by operating as a sequence of passes in which pixels are removed in turn from edges with different orientations.

To design a parallel thinning algorithm, we first construct a mask

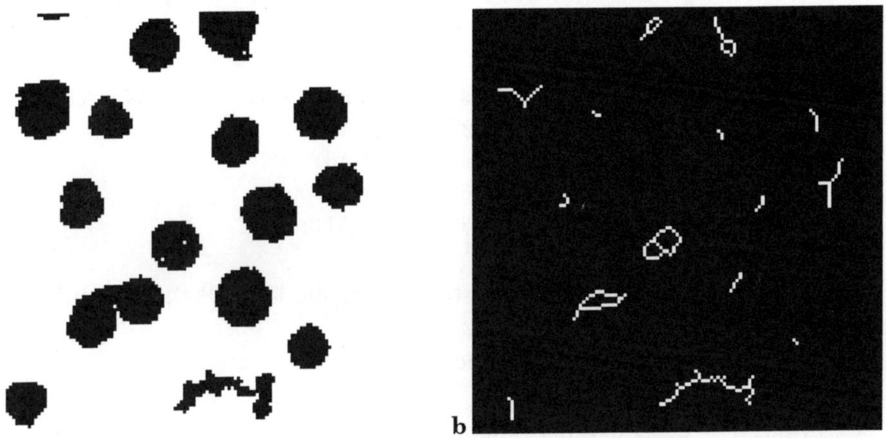

a b

Figure 3.2 Skeleton formation. (a) shows a thresholded version of the peppercorn image of Figure 2.7. (b) shows the result of applying a thinning algorithm to form skeletons of the various objects. Notice how the shape of the twiglet (an example of EVM) is reasonably well represented by its skeleton, and includes a T-junction. Note also that some of the round peppercorns do not have point skeletons because of spurs originating from prominences on the boundaries: in some other cases, single pixel holes have caused closed loops to appear in the skeletons, indicating the dangers intrinsic to this type of representation. See Figure 2.7 for a further example of skeletonization.

which can be used to locate object boundaries with a given nominal direction. Thus we might choose:

$$\begin{array}{ccc} \times & 0 & 0 \\ \times & 1 & 0 \\ 1 & \times & \times \end{array}$$

where the '×' represents an unknown value. Then we set up a test as to whether removing the central 1 might cause disconnection to occur, the crossing number test presented below being the standard one to use for this purpose. Next we generate all other masks we wish to use to help locate suitable edge points of different orientations. Finally, we assemble them into a complete algorithm of the form shown in Table 3.2. The reason for thinning in a *sequence* of orientations is that parts of an object that are two pixels wide could be eliminated if several orientations were tackled simultaneously.

The *crossing number* concept involves counting the number of 0-to-1 and 1-to-0 transitions made as one proceeds once around the current pixel (Table 3.3). The concept is complicated by adherence to 8-connectedness for 1's and 4-connectedness for 0's. Thus an unexpectedly complex formula has to be used:

```
chi := num(A1 ≠ A3) + num(A3 ≠ A5) + num(A5 ≠ A7) + num(A7 ≠ A1)
        + 2*(num(not A1 and A2 and not A3)
          + num(not A3 and A4 and not A5)
          + num(not A5 and A6 and not A7)
          + num(not A7 and A8 and not A1));
```

where the function num(·) is used to convert the logicals 0, 1 to numbers of the same value.

Table 3.2 Basic parallel thinning algorithm. Note that this form of the algorithm could shorten skeleton limbs (see text)

```
repeat
   for each edge orientation do
      for each border pixel do
         if direction mask gives a preliminary match
            then
               begin
                  calculate crossing number chi;
                  if chi = 2 then B0 := 0
                     else B0 := A0;
               end;
until no further change;
```

Table 3.3 Crossing number (χ) configurations. In these instances, 0 is taken to represent the background and 1 the foreground.

0 0 0	0 1 0	0 0 0	1 0 0
0 1 0	1 1 1	0 1 0	1 1 1
0 0 0	0 1 0	1 0 0	1 1 1
$\chi = 0$	$\chi = 0$	$\chi = 2$	$\chi = 2$
0 1 0	1 0 0	1 0 0	1 0 1
0 1 0	1 1 1	0 1 1	0 1 0
1 0 0	1 0 1	1 0 1	1 0 1
$\chi = 4$	$\chi = 4$	$\chi = 6$	$\chi = 8$

There is one final detail lacking in the above algorithm. It is possible that the skeleton will be thinned to single pixel width in some positions before this happens at other positions. In that case a skeleton limb might be shortened by successive pruning operations at its end. Since it is generally desirable to make the shapes of skeletons reflect as accurately as possible the shapes of the objects from which they arise, it is useful to prevent skeleton ends from being pruned in this way. Hence a simple test of the sum of all the pixel values around the periphery of the window should be made, and if it is unity, removal of the central pixel should be inhibited.

Finally, skeletonization of real (i.e. non-idealized) objects frequently results in the production of significant numbers of *noise spurs*, i.e. skeletal lines that do not result from genuine object limbs. Algorithms may attempt to avoid this but run the risk of eliminating genuine limbs. The problem is a fundamental one, and arises since algorithms which operate on purely local data will necessarily find it difficult to make the global decisions necessary to distinguish between noise spurs and genuine limbs. In this respect, algorithms which retain global data are less liable to make erroneous decisions. This explains the interest in skeletonization procedures based on distance functions (see below).

3.4 Skeleton-based analysis of shape

We have already indicated that the skeleton representation can be used for the analysis of shapes. In particular, the numbers and types of nodes on skeletons and the limb lengths and orientations, and also the incidence of loops, all indicate characteristics of shapes and provide measures by which they may be recognized. For example, a letter *T* has one central junction node and three line-end nodes (though the situation is more complex for a font with serifs), while an *O* has a single loop and (in the absence of noise spurs) no line ends. In addition, the skeleton can be augmented with distance function values, which will indicate the local limb widths, thereby providing additional classification information.

To analyse nodes, crossing numbers can be useful, though it has recently been found that a modified form of crossing number based on the number of non-zero pixels around a given node location provides an even better measure of the number of skeleton limbs arriving at a node, especially when the number of limbs is considerable. The difference between the two crossing numbers arises because the usual one (described in the previous section) was designed for the purpose of removing points to form a skeleton, whereas the modified one—which is specific to node analysis—can make use of the additional information that the node is already part of a valid skeleton (Davies, 1997e).

3.5 Distance functions

The distance function of a binary picture object is a numerical map of that object showing how far each pixel is from the nearest background pixel (see Figure 3.3). Object boundary pixels will be assigned the value 1; pixels adjacent to these will be assigned the value 2; pixels adjacent to these will be assigned the value 3; and so on. It clearly makes a difference whether an 8-connected or a 4-connected adjacency measure is used within objects, but here we concentrate on the former.

It is quite simple to devise a parallel algorithm to implement such a distance function. Essentially, it will require N passes over the image where the furthest object pixel from the background is N horizontal, vertical or diagonal units away from the background. Before each pass, a parameter containing the distance function value to be marked on the next set of adjacent pixels must be incremented. An important feature of the algorithm is the need to repeat the process of marking distance function values until no more change occurs in the output image. This procedure is somewhat wasteful in that a final pass is needed which makes no changes but merely observes that nothing has happened: there appears to be no method of overcoming this problem, which is actually quite a common one and also applies to thinning algorithms (see above).

While N passes are needed to implement a distance function using parallel processing, a mere two passes are needed using sequential processing. In the first pass the object pixels are numbered in increasing values from left to right and from top to bottom; and in the second pass the object pixels are numbered in increasing values from right to left and from bottom to top. However, care is needed so that the minimum of the two numbering schemes in employed in the final output image. Clearly, the first pass should be a normal forward raster scan and the second pass a reverse raster scan:

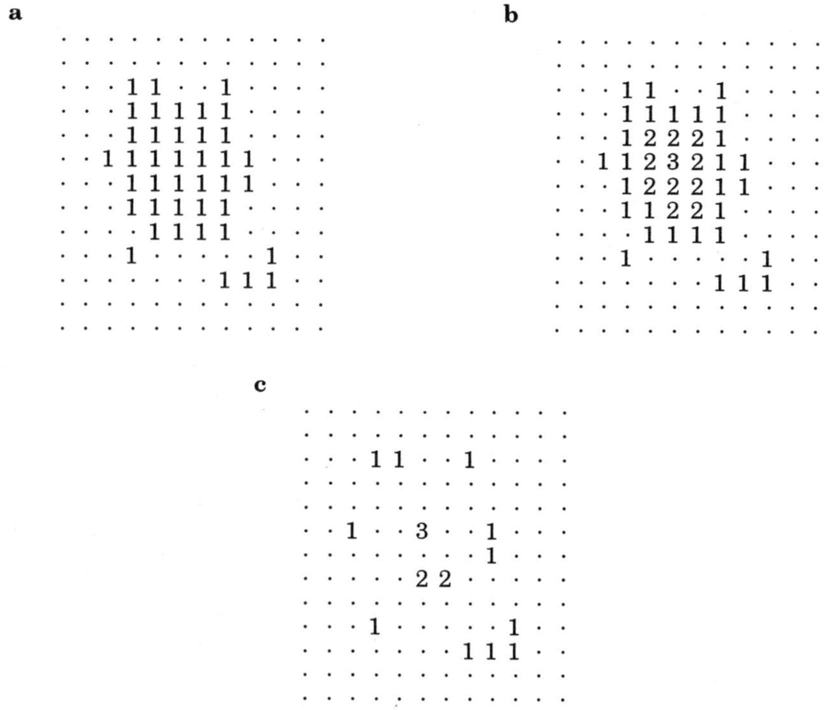

Figure 3.3 Distance function and set of local maxima. (a) shows the initial binary image, (b) shows the distance function, and (c) shows the set of local maxima with values greater than 0. The set of numbered points in (c) is sufficient to reproduce the whole binary image shown in (a).

```
for each pixel in image do
    Q0 := 0;
for each non-border pixel in forward scan direction do
    if A0 > 0 then Q0 := min(Q2, Q3, Q4, Q5) + 1;
for each non-border pixel in reverse scan direction do
    if A0 > 0 then Q0 := min(Q0, Q6 + 1, Q7 + 1, Q8 + 1, Q1 + 1);
```

Once the distance functions have been obtained, their local maxima[2] can be used to gauge the sizes of objects they are in, and to indicate the centre positions of circles, squares and other symmetrical shapes. The local maxima also provide useful markers at near-medial positions within object limbs. Thus they can be of help in providing accurate anchor points for skeletons, though it must be emphasized that the local maxima are not guaranteed to provide a full set of connected links within any object. Indeed, it is possible to prove that such links are disconnected (e.g. a 2

[2]A local maximum is defined as a pixel location whose distance function value is greater than, *or equal to*, that of every one of the surrounding pixels.

cannot be adjacent to a 3, since if it were, it would, by definition, not be a local maximum and would therefore have to be discounted from the set—see Figure 3.3(c)).

It should be remarked that the set of local maxima of the distance functions can in principle be used for a totally separate purpose—that of low redundancy storage and transmission of binary data. A process of *downward propagation* can be used to derive the whole distance function from the set of local maxima; this means that the original shape can be recovered identically from this reduced information set, though the gains will not be great for small images of the type shown in Figure 3.3.

Unfortunately, the simply derived distance functions described above have limited accuracy. This is clear as a local maximum of value N has a footprint (by downward propagation) whose shape is a square of side $2N - 1$, though ideally the footprint should be circular. This deficiency can readily be improved and an octagonal footprint obtained, but further improvements require a significant increase in computation and the simplicity of the basic scheme is lost. In many cases this means that a different approach to the shape analysis task will be needed: because of the slightly limited applicability of the distance function technique, we will not discuss it further here.

3.6 General dilation and erosion operators

In Chapter 2 we encountered the concepts of dilation and erosion. We shall develop these concepts further here as they are important for the analysis and recognition of shapes. One type of dilation operation defined in Section 2.3.2 was that of extending objects to the right. It will be useful to reformulate the operation as:

```
for each non-border pixel do
   B0 := A0 or A5;
```

In this form the operation may be re-expressed in terms of the mask B:

	1	1

where a 1 is placed at the A0 position and the position *opposite* to the A5 position, to emphasize that an extension to the right is intended. Just as it is possible to extend objects to the right, so also it is possible to contract them on the right, using an erosion operation of the form:

```
for each non-border pixel do
    B0 := A0 and A1;
```

In this form the operation may be re-expressed in terms of the mask B':

1	1	

Since this erosion operation cancels out the previous dilation operation in cases where no mergers with nearby objects have been induced by the dilation, it is natural to explain their operation in terms of mask B instead of mask B', noting that when performing an erosion we must first reflect the mask through the origin.

We can develop these ideas to include dilations in which objects are extended in a variety of directions as specified by a given mask: in each case the result for each 1 that is included in the mask will be ***or***ed together to give the overall result of the dilation. In particular, if we extend objects in all directions, as in the second example of Section 2.3.2, the result may be expressed as the following mask C:

1	1	1
1	1	1
1	1	1

Clearly, the corresponding erosion operation employing mask C will also act isotropically. (Notice that, in this case, reflecting the mask through the origin will not modify it.)

Many objects will be restored exactly to their original shapes and sizes on performing a dilation followed by an erosion using the same mask. However, there are two particular situations in which this will not be so:

1. When two objects are very close together and dilation merges them.
2. When parts of the same object are very close together and dilation merges them.

Similarly, performing an erosion and then a dilation using the same mask will tend to produce significant cancellation, which may in certain cases be exact. However, if an object is very thin in some position, eroding it may well split it into several parts, or eliminate part of it altogether, so that a subsequent dilation will be unable to restore it to anything like its original shape. In such cases it is interesting that we can regard erosions as merging two parts of the background that are close together. Thus there is a strong analogy between what erosions do to the foreground and what dilations do to the background. Indeed, the analogy can be formalized as a mathematical dualism between dilation and erosion, as we shall now see.

If a dilation operation represented by mask B is applied to an image A, the result is indicated formally as image $A \oplus B$. Similarly, if an erosion operation represented by mask B is applied to an image A, the result is indicated formally as image $A \ominus B$. Next, let A^c represent the complement of A, and B^r represent the reflection of B in its origin (the latter is at the centre of all the masks presented in this chapter). The dualism between dilation of the foreground (A) and erosion of the background (A^c) can now be expressed by the exact mathematical result:

$$(A \oplus B)^c = A^c \ominus B^r \tag{3.1}$$

while the dualism between erosion of the foreground and dilation of the background can be expressed as:

$$(A \ominus B)^c = A^c \oplus B^r \tag{3.2}$$

3.7 Properties of dilation and erosion operators

Dilation and erosion operators have some very important and useful properties. Also, the fact that these properties are mathematical in nature and are subject to strict mathematical proof brings a certain rigour to the analysis of shape. Hence it is relevant to give a few useful results.

A typical but simple result is the following:

$$A \oplus (B \cup C) = (A \oplus B) \cup (A \oplus C) \tag{3.3}$$

This means that a mask may be split into two separate masks which may then be applied separately and the resulting images ***or***ed together to create the same final image. Such an approach could be useful for providing an efficient implementation, especially in cases where very large masks are involved. For example, we could dilate an image horizontally and vertically by two separate operations which would then be merged together. The following identity shows how an erosion operation could be split in a similar way:

$$A \ominus (B \cup C) = (A \ominus B) \cap (A \ominus C) \tag{3.4}$$

Next, let us replace C in both of these formulae by the identity operation I, which corresponds to a mask with a single 1 at the central (A0) position:

	1	

If we write the union of B and I as D, so that mask D is bound to contain a central 1 (i.e. $D \supseteq I$), we have:

$$A \oplus D = A \oplus (B \cup I) = (A \oplus B) \cup (A \oplus I) = (A \oplus B) \cup A \tag{3.5}$$

which always contains A:

$$A \oplus D \supseteq A \tag{3.6}$$

Similarly:

$$A \ominus D = A \ominus (B \cup I) = (A \ominus B) \cap (A \ominus I) = (A \ominus B) \cap A \tag{3.7}$$

which is always contained within A:

$$A \ominus D \subseteq A \tag{3.8}$$

Operations (such as dilation by a mask containing a central 1) which give outputs that are guaranteed to contain the inputs are termed *extensive*, whilst those (such as erosion by a mask containing a central 1) for which the outputs are guaranteed to be contained by the inputs are termed *antiextensive*. Clearly, extensive operations tend to extend objects and antiextensive operations tend to contract them.

Another important type of operation is the *increasing* operation. An increasing operation is one such as union which preserves order in the size of the objects on which it operates. If object F is small enough to be contained within object G, then applying erosions or dilations will not affect the situation, even though the objects change their sizes and shapes considerably. We can write these conditions in the form:

$$\text{if} \quad F \subseteq G \tag{3.9}$$

$$\text{then} \quad F \oplus B \subseteq G \oplus B \tag{3.10}$$

and $F \ominus B \subseteq G \ominus B$ (3.11)

Interestingly, dilation is associative:

$$(A \oplus B) \oplus C = A \oplus (B \oplus C) \tag{3.12}$$

though the same is not true of erosion:

$$(A \ominus B) \ominus C = A \ominus (B \oplus C) \tag{3.13}$$

However, the fact that:

$$X \oplus Y = Y \oplus X \tag{3.14}$$

means that the order in which dilations of an image are carried out does not matter, and the same applies to the order in which erosions are carried out:

$$(A \oplus B) \oplus C = (A \oplus C) \oplus B \tag{3.15}$$

$$(A \ominus B) \ominus C = (A \ominus C) \ominus B \tag{3.16}$$

Finally, we note that erosion can be used for locating the boundaries of objects in binary images:

$$P = A - (A \ominus B) \tag{3.17}$$

There are many practical applications of dilation and erosion which follow particularly from using them together, as we shall see below.

3.8 Closing and opening

Dilation and erosion are basic operators from which many others can be derived. Earlier, we were interested in the possibility of an erosion cancelling a dilation and vice versa. Hence it is an obvious step to define two new operators which express the degree of cancellation: the first is called *closing* since it often has the effect of closing gaps between objects; the other is called *opening* because it often has the effect of opening gaps (Figure 3.4). Closing and opening are formally defined by the formulae:

$$A \bullet B = (A \oplus B) \ominus B \tag{3.18}$$

a

```
· · · · · · · · · · · ·
· · · · · · · · · · · ·
· · · 1 1 · · 1 · · · ·
· · · 1 1 1 1 1 · · · ·
· · · 1 1 1 1 1 · · · ·
· · 1 1 1 1 1 1 1 · · ·
· · · 1 1 1 1 1 1 · · ·
· · · 1 1 1 1 1 · · · ·
· · · · 1 1 1 1 · · · ·
· · · 1 · · · · · 1 · ·
· · · · · · · 1 1 1 · ·
· · · · · · · · · · · ·
· · · · · · · · · · · ·
```

b

```
· · · · · · · · · · · ·
· · 1 1 1 1 1 1 1 · · ·
· · 1 1 1 1 1 1 1 · · ·
· · 1 1 1 1 1 1 1 · · ·
· 1 1 1 1 1 1 1 1 1 · ·
· 1 1 1 1 1 1 1 1 1 · ·
· 1 1 1 1 1 1 1 1 1 · ·
· · 1 1 1 1 1 1 1 1 · ·
· · 1 1 1 1 1 1 1 1 1 ·
· · 1 1 1 1 1 1 1 1 1 ·
· · 1 1 1 · 1 1 1 1 1 ·
· · · · · · 1 1 1 1 1 ·
· · · · · · · · · · · ·
```

c

```
· · · · · · · · · · · ·
· · · · · · · · · · · ·
· · · · · · · · · · · ·
· · · · · · · · · · · ·
· · · · 1 1 1 · · · · ·
· · · · 1 1 1 · · · · ·
· · · · 1 1 1 · · · · ·
· · · · · 1 1 · · · · ·
· · · · · · · · · · · ·
· · · · · · · · · · · ·
· · · · · · · · · · · ·
· · · · · · · · · · · ·
· · · · · · · · · · · ·
```

d

```
· · · · · · · · · · · ·
· · · · · · · · · · · ·
· · · 1 1 1 1 1 · · · ·
· · · 1 1 1 1 1 · · · ·
· · · 1 1 1 1 1 · · · ·
· · 1 1 1 1 1 1 1 · · ·
· · · 1 1 1 1 1 1 · · ·
· · · 1 1 1 1 1 1 · · ·
· · · 1 1 1 1 1 1 · · ·
· · · 1 · · · 1 1 1 · ·
· · · · · · · 1 1 1 · ·
· · · · · · · · · · · ·
· · · · · · · · · · · ·
```

e

```
· · · · · · · · · · · ·
· · · · · · · · · · · ·
· · · · · · · · · · · ·
· · · 1 1 1 1 1 · · · ·
· · · 1 1 1 1 1 · · · ·
· · · 1 1 1 1 1 · · · ·
· · · 1 1 1 1 1 · · · ·
· · · 1 1 1 1 1 · · · ·
· · · · 1 1 1 1 · · · ·
· · · · · · · · · · · ·
· · · · · · · · · · · ·
· · · · · · · · · · · ·
· · · · · · · · · · · ·
```

Figure 3.4 Results of morphological operations. (a) shows the original image, (b) the dilated image, (c) the eroded image, (d) the closed image, and (e) the opened image.

$$A \circ B = (A \ominus B) \oplus B \tag{3.19}$$

Closing is able to eliminate salt noise, narrow cracks or channels and small holes or concavities[3]. Opening is able to eliminate pepper noise, fine

[3] Here we continue to take the convention that dark objects have become 1's in binary images, and light background or other features have become 0's.

hairs, and small protrusions. Thus these operators are extremely important for practical applications. Furthermore, by subtracting the derived image from the original image, it is possible to locate many sorts of defect, including those cited above as being eliminated by opening and closing: this possibility makes the two operations even more important. For example, we might use the following operation to locate all the fine hairs in an image:

$$Q = A - A \circ B \tag{3.20}$$

This operator and its dual using opening:

$$R = A \bullet B - A \tag{3.21}$$

are extremely important for defect detection tasks.

Closing and opening have the interesting property that they are idempotent: this means that repeated application of either operation has no further effect. (This property contrasts strongly with what happens when dilation and erosion are applied a number of times.) We can write these results formally as follows:

$$(A \bullet B) \bullet B = A \bullet B \tag{3.22}$$

$$(A \circ B) \circ B = A \circ B \tag{3.23}$$

From a practical point of view these properties are to be expected, since any hole or crack that has been filled in remains filled in, and there is no point in repeating the operation. Similarly, once a hair or protrusion has been removed, it cannot again be removed without first recreating it. Not quite so obvious is the fact that the combined closing and opening operation is idempotent:

$$\{[(A \bullet B) \circ C] \bullet B\} \circ C = (A \bullet B) \circ C \tag{3.24}$$

The same applies to the combined opening and closing operation. A simpler result is the following:

$$(A \oplus B) \circ B = (A \oplus B) \tag{3.25}$$

which shows that there is no point in opening with the same mask that has already been used for dilation: essentially, the first dilation produces some effects which are not reversed by the erosion (in the opening operation), and the second dilation then merely reverses the effects of the erosion.

There are a number of other properties of closing and opening; amongst the most important ones are the following set containment properties which apply when $D \supseteq I$:

$$A \oplus D \supseteq A \bullet D \supseteq A \tag{3.26}$$

$$A \ominus D \subseteq A \circ D \subseteq A \tag{3.27}$$

Thus, closing an image will if anything increase the sizes of objects, while opening an image will if anything make objects smaller, though there are clear limits on how much change closing and opening operations can induce.

Finally, it should be noted that closing and opening are subject to the same duality as for dilation and erosion:

$$(A \bullet B)^{\mathrm{c}} = A^{\mathrm{c}} \circ B^{\mathrm{r}} \tag{3.28}$$

$$(A \circ B)^{\mathrm{c}} = A^{\mathrm{c}} \bullet B^{\mathrm{r}} \tag{3.29}$$

3.9 Summary of morphological operations

The past few sections have by no means exhausted the properties of the morphological operations dilate, erode, close and open. However, they have outlined some of their properties and have demonstrated some of the practical results obtained using them. Perhaps the main aim of including the mathematical analysis has been to show that these operations are not *ad hoc*, and that their properties are mathematically provable. Furthermore, the analysis has also indicated (a) how sequences of operations can be devised for a number of eventualities, and (b) how sequences of operations can be analysed to save computation (for instance) by taking care not to use idempotent operations repeatedly, and by breaking masks down into smaller more efficient ones.

Overall, the operations devised here can help to eliminate noise and irrelevant artefacts from images, so as to obtain more accurate recognition of shapes; they can also help to identify defects on objects by locating specific features of interest. In addition, they can perform grouping functions such as locating regions of images where small objects such as lentils may reside. In general, elimination of artefacts is carried out by operations such as closing and opening, while location of such features is carried out by finding how the results of these operations differ from the original image (cf. equations (3.20) and (3.21)); and locating regions where clusters of small objects occur may be achieved by larger scale closing operations. Clearly, care in the choice of scales and mask sizes is of vital importance in the design of complete algorithms for all these tasks. Figures 3.5 and 3.6 illustrate some of these possibilities in the case of a peppercorn image: some of the interest in this image relates to the presence of a twiglet and how it is eliminated from consideration and/or identified.

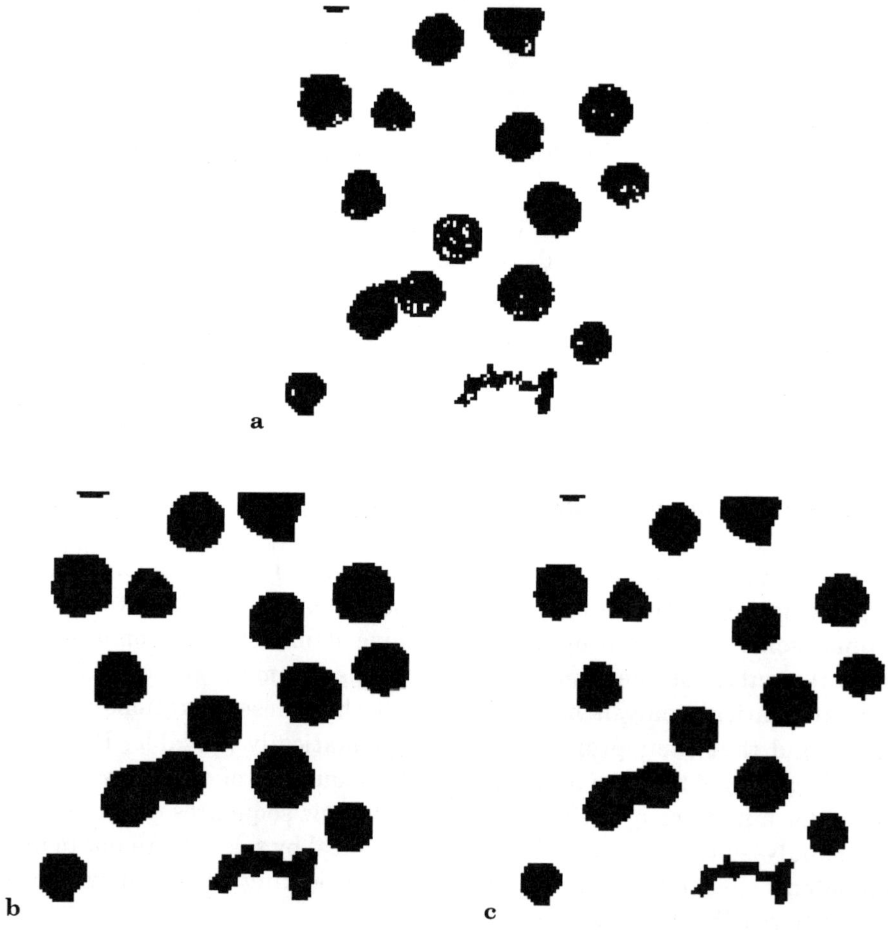

Figure 3.5 Use of the closing operation. (a) shows a thresholded version of the peppercorn image of Figure 2.7. (b) shows the result of applying a 3×3 dilation operation to the object shapes, and (c) shows the effect of subsequently applying a 3×3 erosion operation. The overall effect of the two operations is a 'closing' operation. In this case closing is useful for eliminating the small holes in the objects: this would, for example, be useful for helping to prevent misleading loops from appearing in skeletons. For this picture, extremely large window operations would be required to group peppercorns into regions.

3.10 Boundary pattern analysis

As indicated in Chapter 2, it is possible to analyse digital images by tracking around the boundaries of objects, a process that can be carried out on the original grey-scale image or on the derived binary image. The result of such tracking procedures will be a sequence of boundary coordinates giving an alternative binary shape representation of objects in the image.

This sequence of coordinates can be processed in a number of ways. One simple approach is to deduce the area A and perimeter P of each object and to characterize its shape by finding the dimensionless ratio $C = A/P^2$ (notice that C is independent of the size of the object). Since C is largest for a circle it is frequently called the 'circularity' measure. While C provides a measure of the general compactness of the shape (for which reason it is also often referred to as the 'compactness' measure), it is only one of many such measures, aspect ratio being another.

The result of this approach to shape analysis is a rather *ad hoc* set of parameters which does not guarantee that similar or even totally different shapes can be discriminated from another: in addition, it does not obey the useful maxim that the original shapes can be reproduced, even approximately, from the characterizing parameters. On the other hand, the method of moments does obey this maxim, as it is a systematic, mathematically complete description of the shape; in addition, the infinite series of moment parameters it produces can be curtailed at any desired level of accuracy. However, the fact that it is not especially adapted to any particular type of shape means that a large number of parameters is often required to characterize objects: this contrasts with the skeletonization approach which is well adapted to tasks such as OCR. Space precludes a full analysis of the method of moments and its many variants, but it will be worth considering a simple, intuitive and systematic way of processing object boundary data—the *centroidal profile* approach.

3.10.1 The centroidal profile approach

The centroidal profile is easily computed. First the centroid of the area is determined: alternatively, the centroid of the perimeter pixels may be used. Then the boundary is referred to a polar representation with the centroid as origin, and an (r, θ) plot is drawn representing the boundary shape. From the polar plot, recognition of certain shapes such as circles and regular polygons is straightforward, while other shapes can be recognized with slightly more computation by comparing the polar plot with that contained in a lookup table. Note that the latter will have to be compared in many positions representing the various allowed orientations of the object, though computation can be cut down significantly by a coarse-fine matching

Figure 3.6 Use of the opening operation. (a) shows a thresholded version of the peppercorn image of Figure 2.7. (b) shows the result of applying a 7 × 7 erosion operation to the object shapes, and (c) shows the effect of subsequently applying a 7 × 7 dilation operation. The overall effect of the two operations is an 'opening' operation. In this case opening is useful for eliminating the twiglet. (d) and (e) show the same respective operations when applied within an 11 × 11 window. Here some size filtering of the peppercorns has been achieved and all the peppercorns have been separated—thereby helping with subsequent counting and labelling operations.

approach or by searching for prominent features such as corners (Davies 1997e). The method can also be used for locating any shape defects in an object. However, this technique lacks robustness, since any substantial shape defect will shift the centroid and render the polar plot incomprehensible (Figure 3.7).

d **e**

Figure 3.6 (continued)

It might be thought that shape defects are rare and ought not mitigate against use of an otherwise valuable technique. However, there are several reasons why such a conclusion ought to be considered very carefully:

1. If it is desired to employ the technique for identifying and orientating objects prior to inspection, careful scrutiny will not be possible if an object cannot be identified or orientated: while such objects could be rejected from a product line, this will result in major defects not being properly logged by the inspection system: this is not a desirable outcome.
2. Gross 'defects' can result from two perfectly acceptable objects touching, in which case the tracking routine will continue around both objects, thereby creating an unrecognisable entity.
3. Gross defects can also arise from partial occlusions, when one object is in front of or partially overlapping another. Again, both objects could be entirely acceptable, but neither would be identified by this technique.

For these reasons, in cases where a greater degree of robustness is required, the main cause of the problem—that of reliance on preliminary location of the centroid—will have to be eliminated. Fortunately, a simple route is available to achieve this: use of a representation which uses the boundary to provide its own reference system.

The (s, ψ) plot forms the basis for one such scheme. This explores the orientation ψ of the tangent as a function of the distance s along the boundary from some fixed location on it. However, use of ψ is inconvenient as it increases by 2π every time the boundary is traversed: what is required is some function of arc length which is single-valued. A convenient function of this form is the curvature κ, which is actually the derivative of ψ:

$$\kappa = d\psi/ds \tag{3.30}$$

This function has a constant value for a circle, and has the advantage of being intuitively correct, at least in this important case. For a square, the function has four regions over which it is zero, though there is a slight

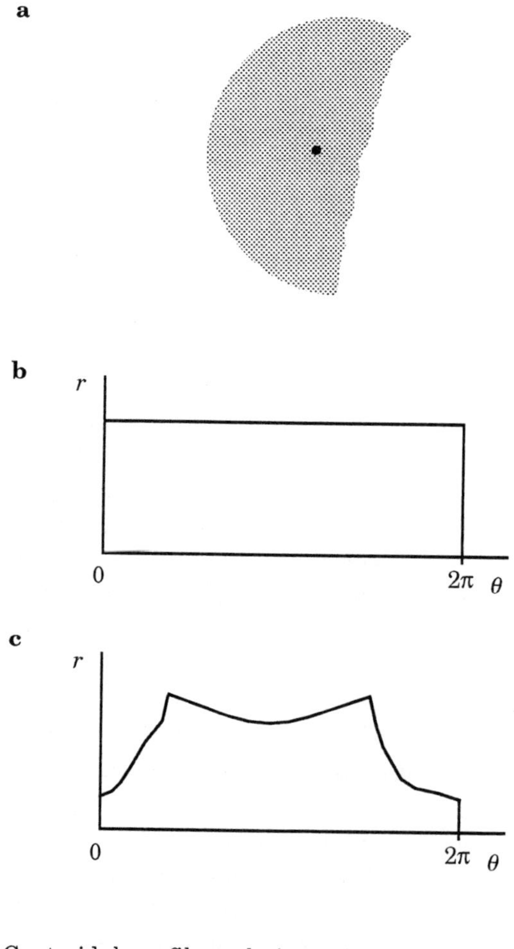

Figure 3.7 Centroidal profile technique. (a) shows a broken circular object, the black dot indicating the position of the centroid; (b) shows the centroidal profile of an ideal round object; while (c) shows the centroidal profile of the broken object presented in (a). Note that the non-broken part of the object boundary in (a) does not lead to a horizontal line in (c) because of the shift of the centroid, the reference point for the computation.

problem in that the curvature is in principle infinite at the corners: nevertheless, when practical means are adopted for measuring the curvature at corners, by integrating the change of orientation over a fixed small distance, this problem is essentially eliminated.

The precise way in which the (s, κ) technique is used for recognition will not be discussed in detail here: neither will the problems involved in using it to solve the problems of occlusion. A full explanation may be found elsewhere (Turney *et al.* 1985; Davies 1997e), but the basic principle is to use it to measure salient features that are then fed to a Hough transform or other indexing technique which can infer the presence of any objects (see Chapter 4).

3.11 Concluding remarks

Binary images contain all the data needed to analyse the shapes, sizes, positions and orientations of objects in two dimensions, and thereby to recognize them and even scrutinize them for defects. Region analysis procedures such as thinning followed by skeleton characterization, and boundary-based procedures such as the centroidal profile technique represent strong traditions within this subject area, and have been applied to very many problems.

In addition, a good many simple small window operations exist for processing binary images and moving towards the same goals. The initially rather random array of such procedures has in the last decade or so been greatly unified by the developing theory of mathematical morphology. As its name suggests, mathematical morphology is mathematical in nature, but this chapter has aimed to provide something of the flavour of the subject. In particular, dilation and erosion have been considered carefully and shown to lead to the key concepts of closing and opening, which are able to filter images in various ways to focus powerfully on relevant types of objects and features.

The concept of connectedness and connected components analysis has been shown to be key to the whole subject and to be highly relevant to the tasks of counting, labelling, separation and containment. Boundary pattern analysis has been considered and found to be problematic in that object defects and distortions can prevent algorithms working well, while difficulties caused by partial occlusions and even touching objects can be insurmountable. It turns out that these problems do not result from applying specifically boundary-based procedures but also occur with region-based analysis. To solve them more advanced techniques are required, such as the inference procedures described in the following chapter.

In the space of a single chapter it is impossible to do justice to all previous work in this crucial area. For a longer but still readable account covering several chapters, the reader is referred to Davies (1997e).

It is interesting to recall that it was Rosenfeld who provided much of the energy underlying the early development of the subject: he perceived the importance of topics such as connectedness and adjacency, and went on to develop the idea of a distance function which is related to the widely used skeleton concept (Rosenfeld and Pfaltz 1966, 1968; Pfaltz and Rosenfeld 1967; Rosenfeld 1970, 1979). In particular, his work laid the foundations for the design of more accurate thinning algorithms: see Arcelli *et al.* (1975, 1981), Davies and Plummer (1981), and Arcelli and di Baja (1985); note that the Davies and Plummer paper includes a rigorous test of the effectiveness of thinning algorithms.

Serra's early (1982) volume on morphology also provided a stimulus which has been maintained to the present day. Haralick *et al.* (1987) generalized the underlying mathematical concepts, while Zhuang and Haralick (1986) developed methods for decomposing morphological operators. The paper by Dougherty and Giardina (1988) was important in the development of methods for grey-scale morphological processing. The reader is referred to the volumes by Dougherty (1992) and Heijmans (1994) for further in-depth information on the subject. A useful tutorial article is provided by Bangham and Marshall (1998).

Chapter 4

Feature detection and object location

4.1 Introduction

Chapter 2 examined basic methods of image processing and found that objects could be located in 2D images by thresholding and adaptive thresholding. Such procedures represent a rather rudimentary and optimistic approach to image segmentation in which it is assumed that objects appear as regions of approximately uniform intensity against an unstructured and uncluttered background—a situation which is generally far from valid in real applications. Although this approach can be coaxed into action in certain constrained environments (e.g. those in factories where the lighting is controlled and the camera is placed in a suitable position near a flat horizontal conveyor), in less artificial situations, more robust methods are required if objects are to be located reliably. In this chapter we consider the problem more carefully, and see how robustness might be achieved in more complex and general environments.

4.2 From template matching to inference

It is first necessary to examine the well-known template matching approach, which corresponds closely with the matched filtering technique used in radar and other 1D signal processing applications (Davies 1993). Figure 4.1 shows the basic situation. A template of the correct shape and intensity profile is moved over the image noting the places where a match occurs. If there are several positions where this happens, it is deduced that an object of the selected type has been located in these positions.

There are several problems with this approach. First, objects are likely to appear in a variety of orientations, so it is possible that (say) 360 templates will be required, corresponding to one per degree of rotation, in order to locate them. The search will therefore take a time proportional both to the size of the image ($N \times N$) and the size of the object (which may be modelled using an $n \times n$ template), giving a total number of operations for a 256×256 image and a 40×40 pixel object of some $64{,}000 \times 1600 \times 360 \approx 36{,}000{,}000{,}000$ operations. This is clearly a prohibitively large number of operations, and is not conducive to real-time applications.

The other main problem with this approach is that it is difficult to define the quality of the match, and an exact match is most unlikely to happen if noisy grey-scale images are involved. With linear systems such as matched filters, the template is convolved with the image—a process that involves multiplying each pixel intensity with a template mask coefficient and summing the results together, obtaining a correlation coefficient. A high correlation coefficient then signifies a match, though 100 % correlation is unlikely, and it is usually left to experimentation to specify a satisfactory degree of correlation. Correlation is a well-attested approach and is often highly successful in signal processing applications, though its success in image processing applications is generally poor if whole objects are to be detected. The reasons for this lie in the huge variability of image signals, through variations in ambient lighting and considerable variations in the objects themselves. For example, objects may exhibit shape variations and also surface texture variations which will prevent them from matching the template at all accurately.

There are two important ways around these problems. One is to employ small masks which look for features rather than whole objects. The other is to give the masks zero mean weighting so that the signals arising from them are unaffected by changes in ambient lighting or similar effects (though the output signal is bound to be proportional to the contrast of the

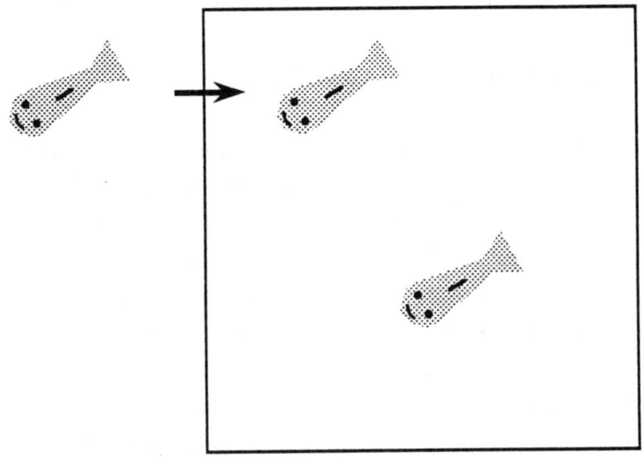

Figure 4.1 Template matching. This diagram illustrates the basic concept of template matching, in which an idealized template for the objects or features to be located is moved systematically over the image.

original signal, which in turn depends on the level of lighting). In this milieu the edge detection approach mentioned at the end of Chapter 2 starts looking quite useful: not only does it employ small template masks but also it is straightforward to find suitable zero mean versions of them (cf. the Sobel masks presented in Section 2.7).

In fact, there is a problem with the strategy of finding objects from their features. As objects are not sought in their entirety, a further stage of processing is required to determine their presence. In the boundary pattern analysis approach, linked sets of edge features were first taken to form complete object boundaries, and the presence and identities of objects then deduced. However, this strategy was not always viable as complete object boundaries could not always be produced. This situation arises in many applications, and very often we have to infer the presence of objects from incomplete evidence—as may happen specifically when parts of boundaries are faint or missing, or parts of objects or boundaries are occluded.

The difference between deduction and inference is that the former refers to logical arguments which arrive at inescapable conclusions, whereas the latter relates to arguments which arrive at logically possible conclusions that are supported by the evidence, while not being the only conclusions which could be compatible with the data. In fact, in image analysis we are looking for the *most likely* such conclusions. The general procedure is to arrive at a number of likely hypotheses and then to check in some way which of them is the most probable. However, it often happens that there is only one reasonable hypothesis and there is no real need to check it—though this does not alter the fact that the process of arriving at a solution in this way is one of inference from incomplete data. In an application involving the inspection of boffles on a boffle line, it is no surprise if an object turns out to be a boffle.

In what follows, we shall first examine various methods of feature detection; we shall then move on to the task of inferring the presence of objects from their features by applying suitable models of the objects that are being sought.

4.3 Finding features

In Chapter 2 we found that edge features could be detected by masks such as the following:

$$\begin{bmatrix} -1 & 0 & 1 \\ -1 & 0 & 1 \\ -1 & 0 & 1 \end{bmatrix}$$

though a complete set of four such masks would be needed to cope with all possible edge orientations. Although the vector approach to edge detection reduced the requirement to just two orthogonal masks, such a solution would probably not be possible in other cases of feature detection. In general, feature detection masks would have to resemble the features that were being sought—reflecting the fact that template matching is a form of matched filtering (Davies 1997e). For corner detection this would mean that masks like the following should in principle be used:

$$\begin{bmatrix} 0 & 0 & 0 \\ 0 & 1 & 1 \\ 0 & 1 & 1 \end{bmatrix}$$

However, the requirement that masks should be zero mean in order to suppress the effects of variations such as those due to ambient lighting necessitates modifying this mask to the following form:

$$\begin{bmatrix} -4 & -4 & -4 \\ -4 & 5 & 5 \\ -4 & 5 & 5 \end{bmatrix}$$

Similarly, a hole detection mask would take a form such as:

$$\begin{bmatrix} -1 & -1 & -1 \\ -1 & 8 & -1 \\ -1 & -1 & -1 \end{bmatrix}$$

while a typical line segment detection mask would have the form:

$$\begin{bmatrix} -1 & 2 & -1 \\ -1 & 2 & -1 \\ -1 & 2 & -1 \end{bmatrix}$$

As in the case of edge detection, there is a problem with the number of masks that are required to cope successfully with all possible orientations of a feature. In general, eight masks will be required in 3×3 windows, though if larger masks were needed to accurately characterize a feature, up to 360 masks might be required as hinted earlier. For corner detection, it seems that eight masks are indeed needed, whereas the number is four for edge detection, four for line segment detection, and one for hole detection. As stated previously, the number drops to two for edge detection, if the vector formulation is used rather than pure template matching. It has recently been shown that this formulation can also be used for line segment detection if a rather unusual set of masks is employed:

$$\mathbf{L}_0 = \begin{bmatrix} 0 & -1 & 0 \\ 1 & 0 & 1 \\ 0 & -1 & 0 \end{bmatrix} \qquad \mathbf{L}_{45} = \begin{bmatrix} -1 & 0 & 1 \\ 0 & 0 & 0 \\ 1 & 0 & -1 \end{bmatrix}$$

These masks give responses g_0, g_{45} leading to:

$$g = (g_0^2 + g_{45}^2)^{1/2} \tag{4.1}$$

$$\theta = \frac{1}{2}\arctan(g_{45}/g_0) \tag{4.2}$$

where the additional factor of one half in the orientation arises as the two masks correspond to basic orientations of 0° and 45° rather than 0° and 90° (Davies 1997d). Nevertheless, these masks can cope with the full complement of orientations as line segments have 180° rotational symmetry and the final orientation need only be given within the range 0° to 180°. For further details of this approach and how it may be applied to the location of insects in grain, see Chapter 10.

Finally, we may note that hole detection masks, which are commonly known as Laplacians, are not only able to detect small dark spots, but are also sensitive to lines. Thus they can be used for line segment detection, though they will have reduced sensitivity in this case.

This seems to complete the possibilities for pure template matching based on the matched filtering concept, at least, within 3 × 3 windows. However, it does not complete all the possibilities for feature detection. In particular, the median filter has been used for corner detection, since median filters tend to round off corners in grey-scale images; this means that subtraction of the filtered image from the original image results in a corner image (Paler *et al.* 1984). Clearly, corners of negative contrast should be presented in a separate image space or problems could arise. A more serious problem is that the 'corner image' will contain quite a lot of noise, as median filters eliminate noise and the image subtraction process will retain the noise in competition with the corners. However, it has been shown that if the corner detection operation is only applied in the vicinity of edges (detected using a Sobel or other operator), then most of the noise is eliminated, and a highly effective corner detector results (Davies 1988d). For large windows this approach results in a reduction in the number of template masks that have to be used and it is computationally more efficient.

An alternative approach to hole detection is provided by the following simple morphological technique: first perform a grey-scale erosion operation and then follow it by subtraction from the original image (see also Chapter 3). What remains in the original image space will then be an image of all the holes in the original image. Again, this procedure will bring out the salt noise as well as the holes, so care will be needed to discriminate between the sizes of the entities produced by this type of algorithm.

These latter two examples show that although convolution masks provide a good starting point for the design of feature detection masks, alternative design strategies may be able to provide improved speed, accuracy, sensitivity, capability for discriminating against noise, and so on. The reader is referred to the literature for further detail on such matters.

We end by pointing out that some features such as edge and line segments are well suited to creating continuous object boundaries or contours, while other features such as holes and corners are better regarded as point features. Certain types of inference algorithm are better suited to handling the former, whilst others are better suited to the latter. Nevertheless, all characteristics of available features can be called into play at the final checking stage, when hypotheses that have been made about the objects that may be present in an image are carefully confirmed: for example, checking on the orientation of corner features may be carried out at this final stage.

4.4 Line location

In this section we consider how the presence of straight lines may be inferred from knowledge of any edge segments that have been detected. It is also possible to infer the presence of straight lines from knowledge of any line segments, but there is so much similarity between the two cases that we ignore the latter one here.

To proceed, we notice that, ideally, any edge segment which lies on a straight line will have the same orientation θ and (when produced) the same distance ρ from the origin in a direction orthogonal to the line (Figure 4.2). This makes it natural to search for relevant edge segments in real images by displaying them in (ρ, θ) space: those segments on the same line will have almost the same values of ρ and θ. In fact, if we merely plot points in (ρ, θ) space, coincidences will be difficult to see. What is required is to find any small regions in (ρ, θ) space where there are lots of points, and if we *accumulate* points in this space, high peaks will stand out and will represent lines in the original image space. Notice that the strategy is to extract relevant data from the image and to transform it to a *parameter space* where line structures are more prominently displayed: not only can a human operator interpret the display, but also the parameter space can be analysed and the lines found automatically by computer.

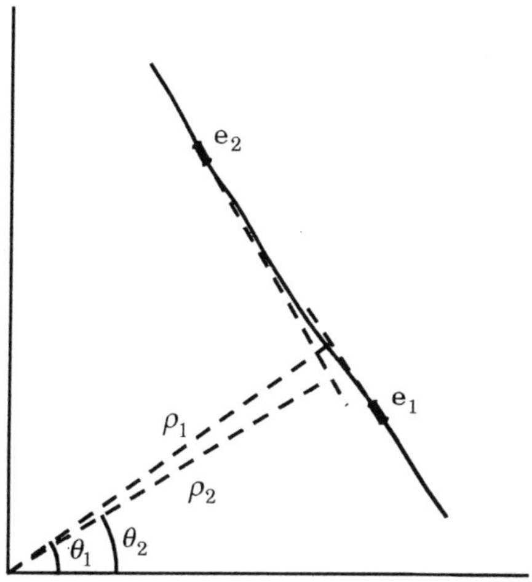

Figure 4.2 Line detection by the Hough transform. Here two edge segments on a line are represented by their normal (ρ, θ) coordinates, ready for accumulation of votes in the (ρ, θ) parameter space.

What is important about this approach is that evidence supporting the existence of straight lines is built up systematically and any missing portions of a line will be ignored. Thus partial occlusions will have little effect on line location, and any parts which are missing for other reasons—such as noise, breakage, distortion, or other effects—will not interfere with the process. Naturally, missing sections of a line will not contribute to the peak that is located in (ρ, θ) space, so sensitivity will be lost, but the result will not be disastrous in the way it is with the centroidal profile approach of Chapter 3.

This *Hough transform* approach has two stages: the first is production of the transform, and the second is location of peaks in the resulting parameter space. So far we have only described the first of these processes. *A priori*, the second process seems entirely straightforward, in that we only have to search for high peaks in the parameter space: applying a threshold should be highly effective in such circumstances. However, peaks are often noisy, and a certain amount of smoothing is frequently necessary so that line parameters may be determined accurately (Davies 1992b). In addition, several peaks including those arising from image clutter may be combined together, so a more intelligent technique involving outlier elimination may be needed. Methods have been devised for achieving this: the relevant subject area is that of robust statistics, though space prevents this topic from being discussed fully here (Davies 1997e). Suffice it to say that peaks can easily appear near to each other, e.g. because of approximately parallel lines due to shadows, and this is a genuine problem.

It should be pointed out that an edge detector with high orientation accuracy is needed to supply appropriate data to a Hough transform: as indicated in Chapter 2, the Sobel operator is useful in this respect, as it is intrinsically accurate to within 1°, while requiring minimal computation by virtue of its small (3 × 3) window. Finally, it should be noted that alternative parametrizations of the Hough transform exist which may be advantageous in different situations—e.g. the slightly faster 'foot-of-normal' parametrization (Davies 1986).

4.5 Circle location

We now consider how the presence of circular objects may be inferred from boundary features, which we again take to be edge segments. As before, it is also possible to start with line segments, with very little change to the basic algorithm, but space prevents us from considering this possibility here.

We proceed to estimate the locations of candidate circle centre locations by moving along each local edge normal through a distance equal to the expected circle radius R, and accumulating a point in a separate image space (Figure 4.3). At this stage the same Hough transform procedure is adopted as for line location, but in this case the parameter space is a real image space which is identical in form to, but distinct from, the original image space. Thus we again search for significant peaks in this space, and take each instance of a high peak as a possible circle centre: note

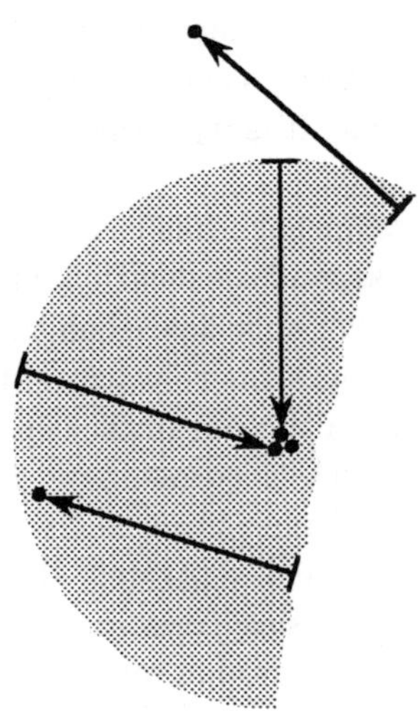

Figure 4.3 Circle detection by the Hough transform. This diagram shows how candidate circle centre locations computed from edge segment positions and orientations are accumulated in the (x, y) parameter space.

that in this case the peaks not only represent the circles but also are situated at their centres, making this a very convenient approach to circle location. Since the peaks only represent possible solutions to the task of circle location, it is, in principle, necessary to confirm them by revisiting the original image data and looking for other information to support the hypotheses that they exist—though in practice, in constrained situations, this stage can often be by-passed (Figure 4.4).

It will immediately be noticed that the above approach to circle location will only work properly if the circle radius R is known in advance. This will not be a major problem in highly constrained situations, such as the location of round cakes. However, if apples are being located the radii may be subject to some variation (as will certainly be the case if the apples are hanging on a tree at variable distances from the camera). In that case the algorithm can be run several times with a range of possible values of R. While this solution might seem to involve a huge amount of computation, this can be cut down significantly by accumulating all the candidate centre locations in a single 2D parameter space, and then locating the highest peak in this space. This approach provides information

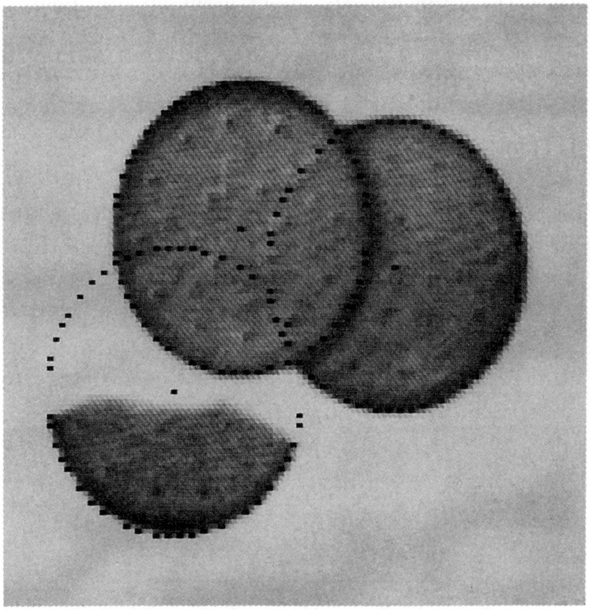

Figure 4.4 Location of broken and overlapping biscuits, showing the robustness of centre location. The accuracy is indicated by the black graphics dots. Note that the lowest biscuit is located slightly less accurately because of asymmetry in the Hough transform peak.

on (a) the number of circular objects, and (b) their positions in 2D. However, it does not determine the radii: these can be estimated by finding which edge pixels contributed to each of the highest peaks and finding their mean distance to those peaks (Davies 1988c). The important thing that is achieved by the main Hough transform procedure leading to the location of the peaks is that of carrying out a rapid search for significant structure in the image. Once this has been done, subsequent analysis of the features in the image becomes far more straightforward and less computation intensive: it is not so much that the Hough transform has achieved, or has failed to quite achieve, the final parametrization of each object, but that it has permitted the computer to key into the image data and effectively eliminate most of it from further detailed consideration.

Numerous other algorithms have been devised for circle location, many of them using the Hough transform approach. In particular, algorithms have been designed which have significantly higher speed or accuracy than the basic approach outlined above: some of these will be considered in Chapter 9 (see also Davies 1987, 1988a,c).

4.6 Ellipse location

The Hough transform approach to object location has been generalized so that any object with a pre-specified shape can be found (Ballard 1981). All that is required is to replace the requirement that we move along each local edge normal through a distance equal to the expected circle radius R by the more general requirement that we move a distance R at an angle ψ to the image x-axis, where R and ψ are functions $R(\theta)$ and $\psi(\theta)$ of the edge orientation θ (Figure 4.5). Indeed, we can restate this requirement more compactly as the need to move by a vector distance $\mathbf{R}(\theta)$ which is a function $\mathbf{R}(R, \psi)$ of the local edge orientation θ. The function $\mathbf{R}(\theta)$ can be

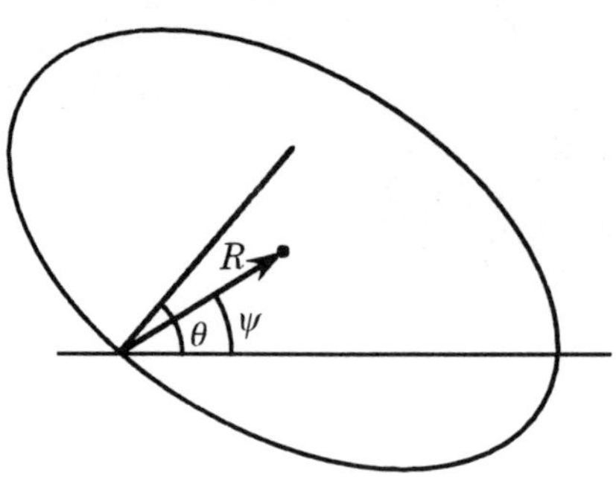

Figure 4.5 Application of the generalized Hough transform for ellipse detection. This diagram shows the geometry for computing the centre location of an ellipse from an edge segment of known orientation.

specified either functionally or in the form of a look-up table, the latter being more convenient in many cases, and also more general in that many shapes are not expressible functionally. In either case it is common to refer to the function as the R-table.

Unfortunately, this generalized approach is complicated by the fact that all shapes other than circles require an additional orientation parameter φ. Hence the R-table should be considered as a whole series of tables corresponding to different values of φ, and this is a distinct encumbrance, though a computationally efficient form has been produced which is capable of detecting ellipses of any orientation (Davies 1989b). However, ellipse location is important both intrinsically and as a means of locating circular objects which are viewed obliquely, so a variety of other approaches to ellipse location have been devised.

Probably the simplest means of ellipse location is the *diameter–bisection* method (Tsuji and Matsumoto 1978). This involves making a list of edge points in an image, and then testing all pairs of points in the list to decide whether they are oppositely orientated. If they are, they could be at opposite ends of ellipse diameters, and the mid-points of the chords joining them are accumulated in a 2D parameter space (Figure 4.6(a)); finally, a normal Hough transform peak search procedure is instigated. The approach works quite well, though for large images with many edge points, the amount of computation can be rather large (with 1000 edge points there will be ~0.5 million points to accumulate in the parameter space). However, the main problem is that the method is not as robust as might be desired,

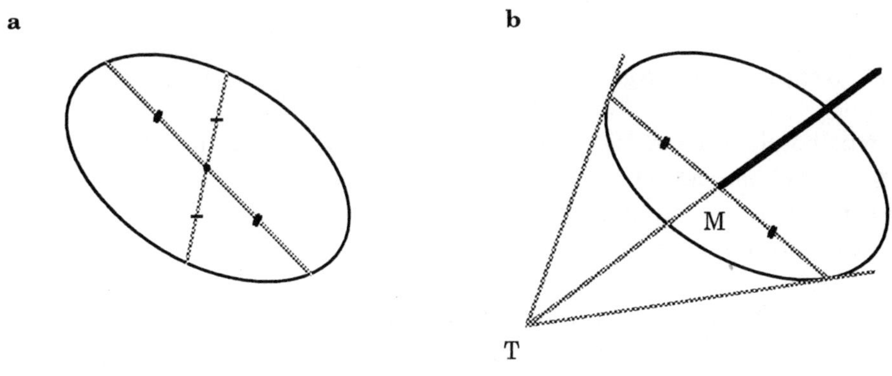

Figure 4.6 Other Hough transform approaches to ellipse detection. (a) shows the *diameter–bisection* method for ellipse detection: the mid-points of chords whose ends have oppositely orientated tangents are accumulated in parameter space. (b) shows the *chord–tangent* method for ellipse detection: points on extended parts of lines (shown thick and dark in the diagram) from tangent junctions (typified by T) to chord mid-points (typified by M) are accumulated in parameter space.

since there is clearly a requirement that there be a sufficient number of whole diameters present—so that any breakages or partial occlusions do not eliminate more than about a quarter of the circumference of the ellipse.

The *chord–tangent* method (Yuen *et al.* 1988) was designed to eliminate this problem. Briefly, it involves taking any pairs of edge points, and constructing the two tangents through them; then drawing a line from the intersection of the tangents T to the mid-point M of the chord between the edge points, and accumulating all the points on the line beyond M (Figure 4.6(b)). It can be shown that the centre of the ellipse must lie on this line, so accumulating all such lines in parameter space will permit the centre of any ellipse to be located. This method is highly robust, but takes significantly more computation than the diameter–bisection method. Means for extracting the ellipse parameters and for cutting down the computation will not be considered here, but it should be noted that the increased complexity arises from the fact that there are five parameters for an ellipse compared with just three for a circle. As a result, it would seem prudent to search quite carefully for other approaches to the location of any objects which possess an even higher degree of complexity. It is noteworthy that considerable effort has been devoted to the location of axes of symmetry in 2D shapes, since these constitute a different way of limiting the search space for complex objects (although we have so far concentrated on the location of object centres in 2D images, the underlying problem is the search for significant peaks in an n-dimensional parameter space, n being equal to five for ellipse location).

Further work on ellipse location is described in Chapter 10 in the context of cereal grain inspection.

4.7 Graph matching

It will be clear from the discussion at the end of the last section that alternative methods which cut down the complexity of the Hough transform are desirable: in particular, it would be better not to be restricted to locating objects from boundary points. Accordingly, we now move on to consider how objects may be located from point features.

In principle, this approach seems far simpler. If we find one point feature of an object, it appears that this will locate the object's position, leaving its orientation unknown: so it is only necessary to find one more point feature to locate the object's position and orientation completely. However, given an object with two features of known separation on an object, if the point features are identical, it will not be known which way up the object is and it will be necessary to find a third feature on it before its location and orientation are defined unambiguously.

Next, there is the problem of robustness against noise, occlusion, possible fragmentation of the object, and a variety of other distortions which could in principle confuse an algorithm. In addition, it should not be forgotten that when point features are involved, inference algorithms will have no *a priori* way of distinguishing such features on one object from those on another. The key to solving this sort of problem is to design algorithms which can cope successfully with the redundancy arising from a number of features: relying on two or three is not a recipe for success.

A simple approach to the problem is to consider the original image as a graph G, and the template of an ideal object as another graph T[1]. What is then needed is an algorithm which searches not only for instances of T within G, but also for instances of part of T within G. We can state this problem as being that of identification of subgraphs of T in subgraphs of G—the so-called *subgraph–subgraph isomorphism* problem. Of course, we could still solve this problem by looking for pairs or triples of features of T in G: this would not make the solution maximally robust. We really need to identify *maximal* subgraphs of T which appear in G (though it is not at this stage clear quite what criteria we should use to identify maximal subgraphs).

To proceed with this task, we can examine all pairs of image features and determine which pairs of template features they could correspond to, taking separation distance D_{ij} as the relevant metric. Thus if D_{ij} in the image equals D_{pq} in the template, we can deduce that $i = p$ and $j = q$ or $i = q$ and $j = p$. For example, if we have a 4-point image (1, 2, 3, 4) and a 3-point template (A, B, C) with $i = 1, j = 2, p = \mathrm{A}, q = \mathrm{B}$, the above relations lead to a compatibility between the feature assignments A1 and B2, or a compatibility between the feature assignments A2 and B1. We express these compatibilities by drawing lines in a match graph, as shown in Figure 4.7. When all possible compatibilities between image and template features have been identified and included in the match graph, it is possible to start inferring the presence of objects in the image. The earlier discussions showed that we want to find maximal subgraphs of T within G. To achieve this we can search for maximal completely connected subgraphs (commonly called *maximal cliques*) within the match graph. In the match graph of Figure 4.7(c), there are four maximal cliques, A1–B2–C3, A2–B1, B3–C2, C1–A3: we interpret this as meaning that the object 1–2–3 in the image matches the template A–B–C exactly, though parts of the object also match parts of the template. In this method of matching it is usual to assume that the largest maximal cliques represent the most likely interpretations of the image features (Bolles and Cain 1982); and once image features have been used in object interpretations, they may not be used in further interpretations. These latter somewhat arbitrary assumptions reflect the fact that the solutions provided by this approach represent inferences rather than deductions: they are considered

---

[1] For the present purpose, a *graph* may be defined as a diagram containing a set of points (often called *vertices*), some of which may be joined by lines (often called *edges*).

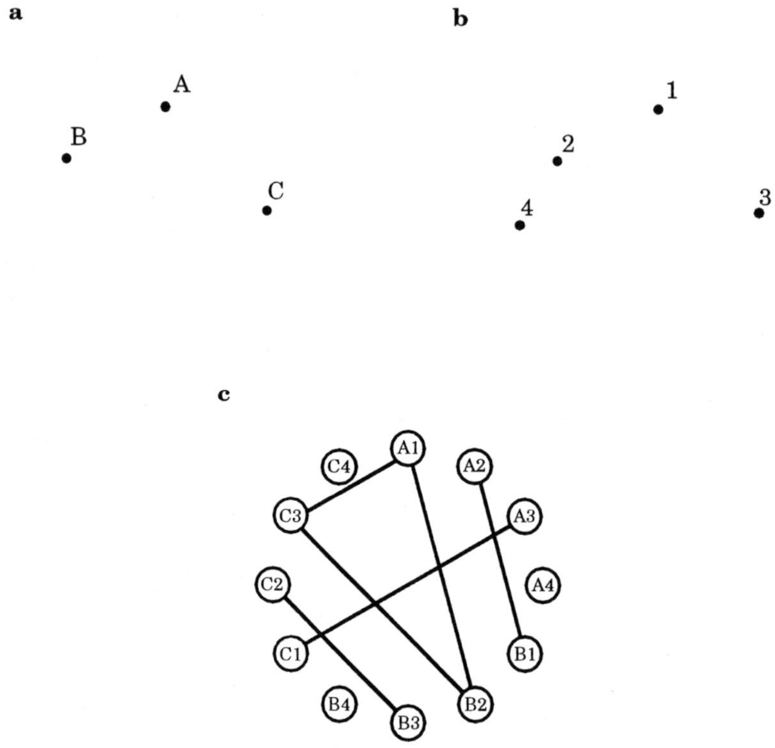

Figure 4.7 The maximal clique approach to object location. (a) shows the template of an idealized object with point features A, B, C; (b) shows the point features 1, 2, 3, 4 actually observed in an image; (c) shows the *match graph* for this image, the most likely interpretation of the data being represented by the largest maximal clique, in this case the triangle A1–B2–C3.

to be amongst the most reasonable interpretations of the image, though there is no certainty that they are in fact correct or even the most probably correct interpretations.

Of rather more consequence than possible errors in the interpretations of maximal clique algorithms is the very considerable amount of computation they involve. For this reason much work has been carried out to find alternative methods of feature-based interpretation. For instance, the maximal clique approach has itself been speeded up (Davies 1991b). Furthermore, it has been emulated by a method related to the generalized Hough transform, as we shall see in the following section.

Before proceeding further, it is as well to notice that the method described above is based on close matching using a direct distance metric. In fact, there are situations where distances may be changed by projection.

We ignore here general 3D projective problems (see Chapter 6), but note that if objects are at an unknown range so that their sizes are all increased or reduced in size by a fixed unknown factor, it should be possible to produce a histogram of all the inter-feature distances in an image; if this is an accumulated histogram (a 1D Hough transform), the characteristic positions of peaks within it should be a scaled version of those for some ideal image containing the objects, and it should be possible to deduce the precise scale factor. Then it should be possible to use the previous matching method to identify the objects in the image.

4.8 Using the Hough transform for point pattern matching

Earlier, it was noted that the Hough transform could be used with pairs of features to locate ellipses, via the diameter–bisection method or the chord–tangent method. In particular, the latter gave a largely unconstrained line of points on which the centre of the object could lie. Thus there was an element of ambiguity in the solution, though this was eventually eliminated by the confluence of lines passing through the centre of the ellipse. However, we can apply such an approach to the location of object centres, or other reference points on them, by taking a pair of point features and deducing where the centre could lie (i.e. inferring its possible positions). Again, the separation of the image features shows which template features could be involved, and then a look-up in a suitable table will show the relative position of the centre, as shown in Figure 4.8 (Davies 1992a). (Clearly, this approach has an affinity with the generalized Hough transform R-table technique, though the matter will not be discussed further here.) Accumulation of the possible centre locations in parameter space permits peak location and leads to the most likely solutions to the object location problem. It can be shown that this method finds identical solutions to those located by the maximal clique method (Davies 1991a). However, it is far faster as its computational load is proportional to the number of pairs of features in image space rather than the number of searches for larger cliques—the latter being exponential in the number of features on each object. (In fact, there are other problems, such as how best to search the parameter space for peaks, how to determine the object orientation, and so on, but these do not alter the fundamental difference between the computational complexities of the two approaches—mainly because the initial unconstrained search is always the process that dominates the computational load.)

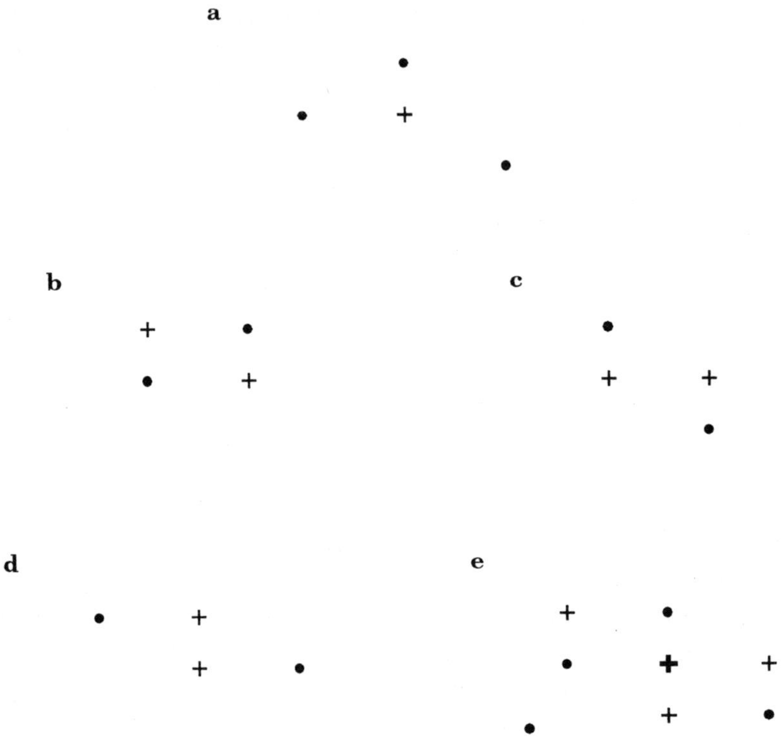

Figure 4.8 Use of the generalized Hough transform for locating objects from point features. (a) shows the template of an idealized object, as in Figure 4.7(a), but with a reference point marked by a cross; (b)–(d) show the position of the reference point relative to the positions of pairs of object features: note that the assumed indistinguishability of the point features leads in all three cases to two possible locations for the reference point, and thus two possible locations of votes in parameter space; (e) shows the positions of votes for the image of Figure 4.7(b): the thickened central cross shows the location with most votes, corresponding to the most likely position of an object reference point.

4.9 Concluding remarks

In this chapter we have seen how the computational load of object location can be cut down from the enormous requirements of the template matching paradigm by taking objects as being composed of easily found features and then employing a suitable feature-based procedure for inferring the presence of the objects. It is important that inference is only capable of producing possible interpretations of any image, though these can be so highly probable as not to need separate checking. Nevertheless, in general unconstrained situations all possible sets of interpretations would need careful comparison, to work out whether some features would be better imagined as part of one object than another—especially in cases where objects are in close proximity or even overlapping. Naturally, it is possible that a feature could be part of one object *or* part of another which it overlaps, and there is no *a priori* method of detecting which interpretation is the more likely; on the other hand, by looking at *all* the evidence in an image, including that from edge *and* point-based features, it may be possible to discern the 'true' situation.

All these factors reflect the following facts: (1) the inference methods outlined here involve somewhat loose assumptions about the most likely solutions; (2) it may not be possible to find the best possible interpretation without appealing to some much larger knowledge base such as that possessed by humans; (3) the inference methods used are *ad hoc* in that the rules they employ seem 'intelligent' in being able to negotiate occlusion and other artefacts, though they do not embody the richness of experience enjoyed by the human. Overall, the approach is model based, and the models involved—especially for the background on which the objects appear—are basic and thereby ultimately failure prone, though this hardly means that they cannot be used in practical situations. Indeed, the Hough transform and maximal clique algorithms described in this chapter provide a very good balance between capability with real data, ease of programming, and computational complexity, and are workhorses for the whole subject area. Table 4.1 summarizes the degrees of robustness that are generally provided by these types of inference algorithm.

Table 4.1 Degrees of robustness

This table expresses degrees of robustness in terms of capability for tolerating various eventualities, such as:

- background clutter
- noisy images
- touching objects
- overlapping and partly occluded objects
- defective objects
- distorted objects.

Chapter 5
Texture

5.1 Introduction

In the context of food, texture means different things to different people. Some regard it as a property of the crunch they get when biting into a piece of cake, apple, meat or celery. Others regard it as the characteristic feel of the surface, for example that of a loaf of bread or a kiwi fruit; others regard it as a characteristic of the way the substance crumbles—as for a cake mixture or a cheese; while others regard the appearance of the surface as the important parameter. Clearly, there is a degree of commonality in all these aspects, and people are judging in their separate ways the surface or bulk structure of the material, the method of assessment depending on whether their current prime interest is in the quality of the raw material, the adequacy of the cooking, the taste, or how attractive the end-product is. A typical dictionary definition of texture will refer to the arrangement of constituent parts or structure of a material: hence what is of prime importance is the nature of the material rather than its 'feel', taste or appearance. Nevertheless, in machine vision and image processing it is the surface appearance that is normally measured and which is taken to be the fundamentally important characteristic. In what follows this is the working assumption that we shall make: if food quality is to be assessed, it should be remembered that the internal food structure may have to be deduced in some way from the appearance, and the method of achieving this will not be known *a priori*. The main exception to this is when X-ray imaging is involved, in which case the textural image is a representation of the bulk structure of the material. However, we shall defer consideration of X-ray imaging until Chapter 11.

In image processing, texture analysis has long been applied to the study of all sorts of materials and surfaces. Materials as varied as sand, grain, grass, water, basketware, linen, stone, paper, wallpaper patterns, bread and other food commodities can all be described as possessing texture. Here texture is taken as a variegated surface: such a surface may either be rough or appear to be rough—as in the case of some wallpaper patterns. But in the image the texture is manifest as a characteristic variation in intensity or colour. In this monograph we shall not consider colour variations but will concentrate on intensity variations.

Slowly varying intensities are not normally considered to represent textures: in fact, textures are taken to exist in regions where the intensity is rapidly varying or 'busy'. In particular, textures may consist of regular or irregular variations in intensity, though regular variations in biological materials (such as muscle fibres or corn kernels on a cob) are seldom perfectly regular, while irregular variations (e.g. in the placement of grains of sand) are seldom completely random. Indeed, it is the very juxtaposition of regularity and randomness that is the essence of texture. And individual examples of texture all have their own characteristic mixtures of regularity and randomness which must be measured before a texture can be recognized (Figure 5.1). Often, textures arise from large numbers of grains, strands or elements which are very similar in size, shape and sometimes direction, but which vary minutely so that absolute regularity is virtually impossible. Nevertheless, short range order can arise, and it is therefore possible in some cases to characterize a texture by some sort of numerical periodicity measure which is related to the size of the individual textural elements or *texels*.

By now it will be evident that texture is a multi-parameter entity which is difficult to measure, not least because it embodies a degree of randomness which is integrated in an *a priori* unknown way with various degrees of regularity: this sets texture apart from other measures such as size and shape which are in principle subject to exact characterization and assessment. In spite of these difficulties, simple measures of texture exist, and the problem is more of knowing which one to use in a given application than of not knowing how to begin. Before proceeding, it is worth pointing out that texture analysis has two main purposes: one is to recognize texture and particularly to ensure that material surfaces are of adequate quality; the other is to determine where one texture ends and another begins, as part of a process of segmentation in a textured environment.

5.2 Tackling the problem of texture analysis

Because of the near-periodic intensity variation in some textures, many workers felt impelled to try Fourier based approaches to texture analysis, while others tried methods based on autocorrelation functions. In spite of the high degree of periodicity of many fabrics, such approaches were not especially successful, and in less appropriate cases, where the texture is more random than regular, relatively little was achieved. Other researches during the early 1970s resulted in the co-occurrence matrix approach becoming the standard one for texture analysis, though by now it has been largely superseded by alternative techniques.

Figure 5.1 Examples of a variety of textures: (a) broccoli flowers, (b) bubbles in bread, (c) gravy granules, (d) butter beans, (e) lentils, (f) rice, (g) shredded wheat cereal, (h) hessian. Notice how they combine randomness with a degree of regularity, and in the last three cases, a steadily increasing directionality. When the component objects are relatively large, it may be more relevant to consider the scene as a set of objects rather than as a texture.

The idea behind the grey-level co-occurrence matrix approach is to ascertain the frequencies or probabilities with which different pairs of intensities appear throughout a texture if samples are taken in specific relative positions. In particular, if pairs of pixels are selected which have separation $\mathbf{d}$, the joint probability that respective intensity values i and j will arise is taken to be $P(i, j|\mathbf{d})$. For a given sample of the texture, tables of frequencies are computed and the joint probabilities are obtained for a variety of values of $\mathbf{d}$. Clearly, the joint probability tables can be voluminous, as all 256 intensity values at one location can in principle appear in conjunction with all 256 intensity values at the other location. In fact, it is normally appropriate to express any image characteristic such as shape in terms of a handful of parameters, and texture should not be very much different in this respect. Hence it is necessary to condense the large number of parameters in the co-occurrence matrices down to a much smaller number. Of course, some values of $\mathbf{d}$ will be far more appropriate for representing the directionality of the texture and the scale of the texels, so many tables can be eliminated. In addition, the quantization of the intensity indices in the tables can be increased to reduce their size, but the number of parameters will still be very large. It should also be noted that the co-occurrence matrices do not themselves lead to texture recognition, but only provide an initial parametrization. Details of how Haralick and others have extracted 'energy', 'entropy', 'homogeneity', 'inertia' and other parameters by means of relatively *ad hoc* summations of matrix coefficients are beyond the scope of this chapter (Haralick *et al.* 1973; Haralick 1979; Conners and Harlow 1980). Suffice it to say that these derived coefficients have been found capable of texture segmentation and recognition tasks in a variety of applications, though it is not at all clear that the methodology for setting up the relevant classifiers is optimal. Hence many workers have considered a great variety of other approaches which might more obviously lead to optimal solutions. For this reason we shall not pursue the co-occurrence matrix approach further here.

5.3 Laws' approach

Laws' approach to texture analysis appeared at the end of the 1970s and is in many ways far more intuitive and straightforward to optimize than the co-occurrence matrix approach (Laws 1979, 1980). For that reason it will be worth considering it here in some detail.

The basic idea is to apply a range of convolution filter masks that are capable of enhancing various types of image detail which could be useful for indicating the nature of any texture. Each filter leads to a separate processed image, and if any one of these exhibits a high *texture energy*, it is clearly of value for signalling the presence of the texture. In fact, what is relevant is the average energy over a reasonable region corresponding to a number of texels: i.e. it is the *macrostructure* rather than the *microstructure* that constitutes the actual texture. As a result of these considerations, the Laws' approach includes a smoothing operation in each of the image filter channels. The information from the various smoothed images is eventually pooled and fed to a classifier which is used to determine the texture at every pixel location.

Typical convolution filters used in Laws' approach include edge, Laplacian, and averaging filters. For example, a region with a high degree of edginess should be recognizable not only as a highly textured region, but as one with a high directionality; similarly, one with a high Laplacian content would constitute a region with a highly 'grainy' texture. While this aspect of Laws' technique is intuitive, the method is thorough in that all possible convolution filters for 3 × 3 and larger windows are tested for their texture characterization capabilities. In this respect it should be noted that nine linearly independent filter masks can be constructed within a 3 × 3 window; 25 linearly independent masks can be constructed within a 5 × 5 window; and so on. Laws devised systematic ways of constructing all possible masks of this type from a basic set of three 1 × 3 masks:

$$\mathrm{E3} = \begin{bmatrix} -1 & 0 & 1 \end{bmatrix} \tag{5.1}$$

$$\mathrm{S3} = \begin{bmatrix} -1 & 2 & -1 \end{bmatrix} \tag{5.2}$$

$$\mathrm{L3} = \begin{bmatrix} 1 & 2 & 1 \end{bmatrix} \tag{5.3}$$

the basic procedure being illustrated by the following equations, '⊗' denoting a convolution:

$$\mathrm{L5} = \begin{bmatrix} 1 & 2 & 1 \end{bmatrix} \otimes \begin{bmatrix} 1 & 2 & 1 \end{bmatrix} = \begin{bmatrix} 1 & 4 & 6 & 4 & 1 \end{bmatrix} \tag{5.4}$$

$$\begin{bmatrix} -1 \\ 2 \\ -1 \end{bmatrix} \begin{bmatrix} 1 & 2 & 1 \end{bmatrix} = \begin{bmatrix} -1 & -2 & -1 \\ 2 & 4 & 2 \\ -1 & -2 & -1 \end{bmatrix} \tag{5.5}$$

Masks such as L3 and L5, and those derived from them using the above two procedures, have no negative coefficients and give results which depend more on intensity than an intensity variation; hence they are less appropriate for texture analysis. However, all other masks are sensitive to 'edges', 'lines', 'spots', 'waves', 'ripples' and other intensity variations which are characteristic of textures.

Once the filtered images have been produced, they are converted to texture energy images by taking local magnitudes of the output values; for this purpose either absolute or squared magnitudes can be used, the latter corresponding more closely to texture energies, but the former requiring less computation and providing a sufficiently good representation for practical applications. Laws used 3 × 3 masks in his filters and followed these with 15 × 15 smoothing masks to convert the micro-features to macro-features. He then combined the outputs from the various filter channels by feeding them to a classifier to determine the type of texture at each pixel location.

The method has been found highly effective for texture classification by Laws and many other workers, and its intuitive approach is conducive to systematic setup of the relevant parameters. At least one study (Pietikäinen *et al.* 1983) has found the method to be more powerful than the co-occurrence matrix method. If it is more powerful, as seems likely, this must be because it is straightforward to design and employ a mathematically complete set of masks. However, for such a set to be truly complete, it must include masks of all relevant sizes, and the set of masks that should be employed is not necessarily obvious without significant experimentation. Thus in practice there is some room for intuition in the design of a complete texture recognition system based on the Laws' paradigm.

Finally, it should be noticed that the Laws' approach is conducive to highly efficient hardware implementations, as simple convolution masks are involved, and these are followed by squaring or absolute magnitude computations and then by smoothing convolutions; in addition, although the masks for the smoothing convolutions are large, they can be implemented as two 1D convolutions (following the reverse of the type of operation indicated in equation (5.5)), thereby cutting computational load by as much as an order of magnitude.

5.4 Ade's approach

Ade's (1983) approach to texture analysis was essentially identical to Laws' approach, but involved choosing the filter masks rather more carefully following a systematic 'principal components' analysis of the input textural data (Chapter 7). By tuning the masks to the data, it was no longer necessary to use a complete set of filter masks, but rather to concentrate on those which corresponded to the texture components having the most energy.

The method involved taking all the pairs of pixels in a 9-element 3 × 3 window and forming a 9 × 9 covariance matrix **R** for the observed intensities within a textured region. **R** was then diagonalized and the eigenvectors determined. Each 9-coefficient eigenvector corresponded to the nine coefficients of a 3 × 3 convolution filter mask, and the corresponding eigenvalue gave that part of the variance of the original image data that can be extracted by each filter. Thus not only was the texture characterized by a set of variances but also the particular filter masks leading to these variances were identified. This meant that the list of filter masks could be curtailed at any convenient stage when the texture energy was felt to be represented sufficiently well: this is consistent with the usual purpose of principal components analysis (Chapter 7)—that the most significant variations are explained by the largest principal components, with the remainder being taken to be of minimal importance and omitted from further consideration, thereby saving computation.

It is important to note that the covariance matrix **R** is a real symmetric matrix, so its energy eigenvalues are real and positive, and the eigenvectors are mutually orthogonal. As indicated above, the strongest eigenvalues should correspond to the most significant textural detail in the image, and should therefore reflect the directionality and shape of any texels it contains. In addition, if large enough windows are used to construct **R**, they should contain all the information needed to produce optimal masks which take full account of the sizes of the texels. Thus small texels will be characterized by masks with small coefficients around the boundaries of the window. Although this neatly helps to eliminate intuitive aspects of mask design indicated earlier, it may be difficult to implement it properly as the diagonalization procedure will have to be performed with very large covariance matrices. Nevertheless, this possibility indicates quite forcefully the power and systematic nature of the Ade approach to texture analysis.

The Ade method described above was devised for optimal classification of textures. However, an enhanced version of the technique has more recently been devised for optimal discrimination between two textures (Unser 1986). This involves the simultaneous diagonalization of two covariance matrices, and as such is considerably more complex. Unser and Eden (1989, 1990) also developed the method further by specifying more clearly quite what types of non-linearity should be included with the various linear filters in the overall design: the underlying idea was to provide a genuine r.m.s. texture signal at the output of each of the linear filter channels, so as to optimize the output for the final classifier. For further details the reader is referred to the literature.

5.5 Concluding remarks

In this chapter we have seen that the analysis of textures presents significant difficulty, largely because textures take the form of complex statistical variations in intensity which are only partly regular in nature. Early experimentation led to the development of the co-occurrence matrix as a basis for characterizing textures, but this approach tended to involve far more characterizing data than the original images, and significant compaction of the data was needed before they could be used successfully: although they have been applied in real applications, it is not at all obvious how best to compress the matrix data, and it is by no means clear how this can be achieved fully automatically in any new application. These considerations led to the development of Laws' method, which is more straightforward to set up, though for a time it was not known how to optimize the process. However, Ade's approach, which involved principal components analysis, largely overcame this latter problem; it has since been widely used and has been developed further, notably by Unser and his co-workers.

Other workers have taken different routes to the replacement of co-occurrence matrices. Harwood *et al.* (1985) developed the *local rank correlation* method; Vistnes (1989) developed the *forced-choice* method which was aimed particularly at improving the accuracy with which edges can be located between textures, and at eliminating the tendency of Laws' method to miss larger scale textural structures and to blur the texture feature values across edges. Hsiao and Sawchuk (1989, 1990) also aimed to eliminate the latter problem—and the problem that classifiers sometimes find additional texture classes in edge regions—by applying an improved technique for feature smoothing and using probabilistic relaxation to obtain improved judgements of texture regions. Space prevents these and many other methods (such as Markov approaches) to be covered in detail in this monograph.

For further details on the subject the reader is referred to the many reviews which have appeared in the literature over the past few years (e.g. Haralick 1979; Van Gool *et al.* 1985; Reed and Du Buf 1993; see also Davies 1997e). Finally, the reader is referred to Chapter 7 for further discussion of principal components analysis, and Chapter 11 for case studies on the application of texture analysis to X-ray images.

Chapter 6
Three-dimensional processing

6.1 Introduction

In the previous chapters we have considered many aspects of image processing, and these have for the most part been tied firmly to interpretation of individual images which have been taken to represent the 'ground truth' in one way or another. Indeed, the archetypal situation has been a camera looking straight down on a horizontal worktable on which a number of fairly flat objects such as biscuits are lying. However, in the real world, objects are rarely flat and in any case they are often situated in complex spatial arrangements relative to one another. Clearly, the world and the objects in it are decidedly three dimensional, and realistic image processing algorithms have to be designed to cope with this fact: even in food applications, the world of the conveyor is often replaced with the outdoor world in which fields must be ploughed, animals must be counted, apples must be picked, potatoes must be sorted and fish must be sized even as they swim.

Unfortunately, the additional dimension brings with it considerable additional complexity, though on occasion we have had some indication of this through occluding, overlapping and touching objects. Not only will the number of parameters required to specify the position of an object have increased from two to three, but also the number of parameters needed to specify its orientation will have increased from one to three. In addition, the dependency on these parameters is essentially exponential, so that if each is quantized into just 100 levels, the number of configurations per object will have increased from a million or so to around a million million.

There are further problems which arise from restrictions in our senses, and those we can provide computers with. Whereas one eye is sufficient to gather all necessary 2D information, two are less than sufficient to gather the right amounts of 3D data: the additional sense of depth provided by two eyes is tolerable at close range but minimal above

100 metres or so. In any case we only see objects from one side: the equivalent of a 2D image for a 3D scene would involve a good many cameras working in concert to build up the right quantities of global information. With such restrictions, there is considerably increased need for a database of relevant information and knowledge of *a priori* probabilities, coupled with the necessity for inference rather than straight deduction, to provide full 3D interpretation.

Nevertheless, the eye possesses remarkable facility for making sense of 3D scenes—so much so that we might suspect that the complications outlined above are more academic than real. However, their reality soon becomes manifest when we try to write computer algorithms to interpret 3D images. In fact, the apparent ease with which the eye operates is now known to hide a huge amount of processing and many interwoven processing methods. It obtains shape information from contours, stereo, shading, colour, texture, focus, motion, and other cues. Perhaps more important, it is able to integrate the information seamlessly, rapidly, and without conscious effort. In one short chapter, it will be difficult to do more than scratch the surface of the problem: it will only be possible to give an impression of the algorithms that are required for this purpose, and to gain insight into some of the main principles.

6.2 Stereo vision

Perhaps the most obvious feature of the human visual system is the fact that it is binocular. The two eyes look at any scene from two positions and acquire slightly different images. A feature that is at one position in one image will be at a slightly different position in the other image: the difference in position is called the *disparity* D and gives important information on the depth Z of the feature in the scene. For an image acquisition system which is set up as in Figure 6.1, it will be possible to deduce the depths of observed features from their disparity measures, using the following very simple formula:

$$Z = fb/D \tag{6.1}$$

f being the common focal length of the camera lenses, and b being their separation.

An alternative method of estimating depth is to mimic the eye's capability for *vergence* by rotating the cameras through opposite angles α until they point towards a particular feature (Figure 6.2). The depth Z is then given by:

$$Z = \tfrac{1}{2} b \cot \alpha \tag{6.2}$$

In fact, the eye employs vergence *and* measures disparity, and is clearly able to make sense of the resulting data. However, machine vision systems generally concentrate on disparity measurement. It is also possible to measure depth by determining the focusing adjustment required for each object, and this has also been emulated in machine vision systems, though it is not widely used: beyond noting that testing for optimum focus can be achieved by maximizing the local high frequency content in the image, we ignore this approach in what follows.

Application of equation (6.1) would appear to be extremely simple, but in practice there is a severe complication in that each feature on a real-world object has to be identified unambiguously in the two input images. This is more difficult than might be imagined, as the two input images are necessarily different: they are actually distorted versions of each other. There are three aspects of this: one is that there are some features that appear in one image which are not visible in the other image, as the

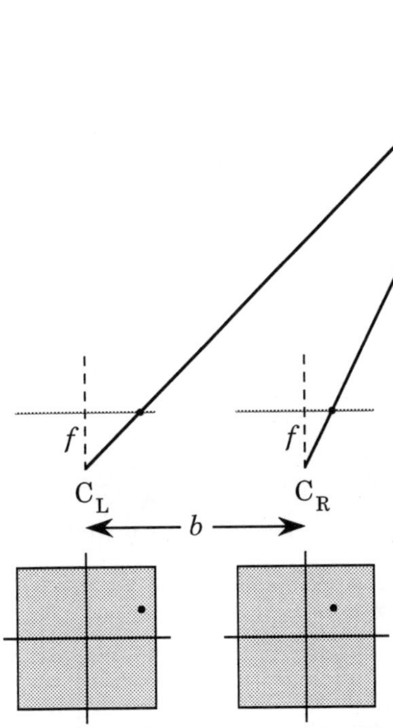

Figure 6.1 Geometry for binocular vision. This diagram shows how an object or feature in a scene is presented in binocular images. C_L and C_R are the *centres of projection* of the two lenses: for convenience of presentation, it is assumed that the images are formed in focal planes a distance f in front of the lenses rather than a distance f behind them. The disparity D between the positions of the feature in the two images allows its depth in the scene to be estimated (see text).

viewpoints are different in the two cases; another is that corresponding features will appear at different positions in the two images; and another is that corresponding features will have different apparent shapes and sizes, so that it may be difficult to tell that they originate from the same part of the same object.

Thus the infamous *correspondence problem* arises—which results in the need to search in each image for features which have been found in the other. An unconstrained 2D search takes a considerable amount of computation, not least because (as already noted) corresponding features will appear slightly different in the two images. Hence it is important to employ some systematic means of cutting down the search. The *epipolar line* approach is the one most commonly used for this purpose. This involves identifying a feature in one image, computing the line of possible positions of this feature in 3D space, and then computing the line of possible image positions in the second image (Figure 6.3). Once this has been done, it is only necessary to do a 1D search along this line (the epipolar line).

While determining the correspondences between features in the two images, there are various factors which can be of help. In particular,

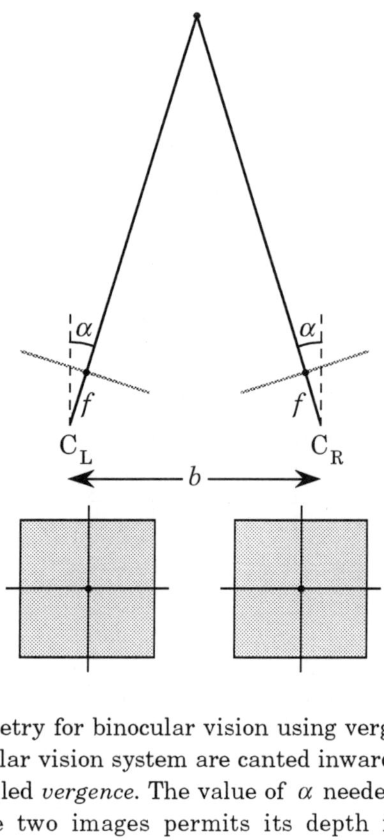

Figure 6.2 Geometry for binocular vision using vergence. Here the two lenses in a binocular vision system are canted inwards through an angle α in a process called *vergence*. The value of α needed to make a feature coincident in the two images permits its depth in the scene to be estimated (see text).

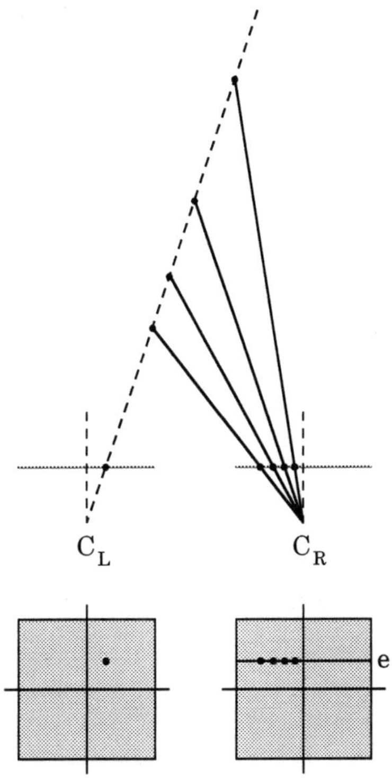

Figure 6.3 Application of the epipolar line concept. Here a feature in the left-hand image may have originated from any point along the dotted line in the scene. This means that the search for corresponding features in the right-hand image can be restricted to the epipolar line e: that this is indeed a line is clear as it is the intersection of the right-hand image plane with the plane containing C_R and the dotted line in the scene.

features on a smooth convex surface will be in the same order in the two images (Figure 6.4): features only appear in a *different* order when they are on different surfaces at different depths in a scene (as when a car is viewed between trees at the edge of a forest); when a feature appears within a semi-transparent object or through a hole in another object; or when features are mis-identified. Thus a global interpretation is more likely to be correct than one which gives answers for a restricted part of a scene.

Clearly, binocular vision should be relatively straightforward when viewing objects possessing easily identifiable features (though if there are too many such features, distinguishing them unambiguously becomes more difficult and computation intensive). It will also be clear that plain feature-free surfaces will be difficult to map in 3D, as there will be very little for the vision system to lock on to. However, surface texture can provide a plethora of useful features, even if they are on an otherwise flat feature-free surface. Thus, anything that will change local surface brightness can be of value in providing matchable features: it is not necessary for the features to originate from exotic 3D shapes or structures. This observation leads to another approach to the production of depth maps of scenes—that of structured lighting.

Structured lighting involves the projection of light patterns onto object surfaces, thereby making otherwise indistinguishable points distinguishable, and producing artificial features on an otherwise plain surface (Figure 6.5). It is especially useful for complementing the usual binocular vision approach, but runs into problems when the surface has strong reflectance variations, which can cause the normal and artificial features to give rise to confusing patterns. The basic idea of structured lighting can be seen by considering the projection of a dot pattern. If a dot is projected from a known angle and observed in an image from another angle, it will be clear that its (reflected) position on an object can be computed, and the depth in the scene estimated. Indeed, the geometry is the same as that for binocular vision, the difference being that the light is projected along the incident line rather than being received by a separate camera along the same line.

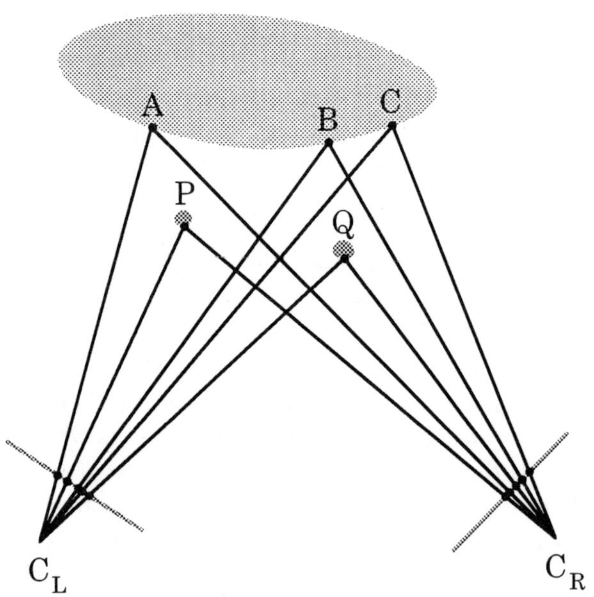

Figure 6.4 Use of information on the ordering of points in binocular images. Here a scene contains three convex objects, the first of which has three prominent ordered feature points A, B, C. Points A, B, C appear in the same order in the two images: the changing order of points P and Q indicates that these lie on objects in front of the object containing A, B, C. The available information is consistent with P and Q lying on separate objects or on a single object containing a hole.

A correspondence problem also arises with structured lighting: all the feature dots must be coded so that they are distinguishable and easily identified in the received image. One way of helping with this problem is to project patterns of dots in which the patterns are easily identifiable: however, though this can be achieved quite readily on a plain flat surface, on a curved surface the patterns become distorted and more effort has to be devoted to decoding; also, ambiguities can creep in, especially near prominences or holes. Various workers have tried light stripes, grid patterns, wide stripes, coloured stripes, and also temporal coding approaches. It is common to use a single light stripe, and to move it over the surface of an object, though this form of temporal coding tends to be slow in operation. An alternative approach is to use a CCD projector to apply light patterns which change dynamically, perhaps the most notable method being to apply a coarse pattern at first, followed by progressively finer patterns until sufficient accuracy is obtained. In view of the huge

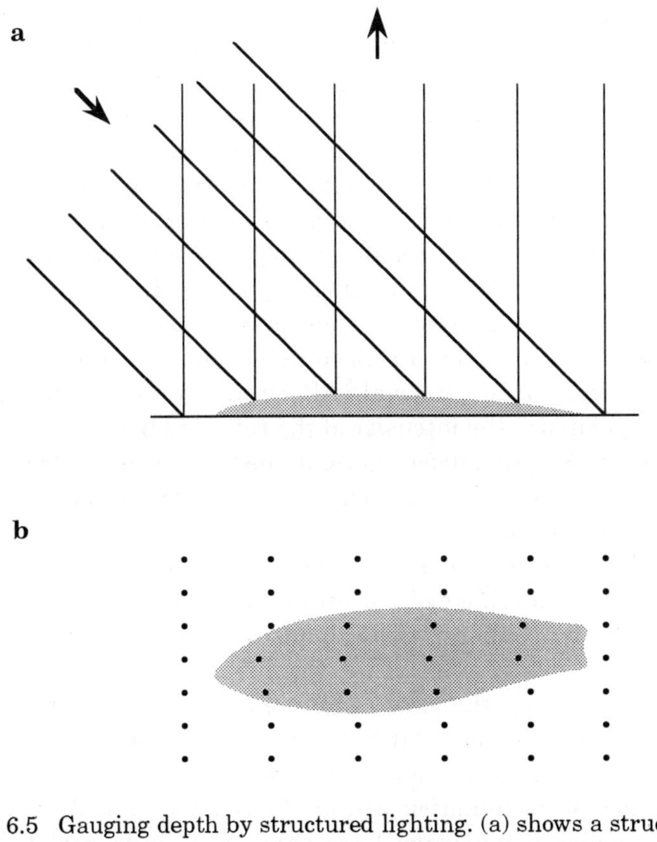

Figure 6.5 Gauging depth by structured lighting. (a) shows a structured lighting scheme, in which rows of dots are projected onto a fish lying on a worktable. (b) shows the lateral displacement of the dots, permitting the thickness of the fish to be estimated at a number of positions. In real applications, the dots are points of bright illumination.

number of possibilities, further discussion of this approach is curtailed.

6.3 Shape from shading

There are many situations in which the eye can discern depth when binocular vision is ineffective and structured light (in the earlier sense) is not applied. This occurs particularly with distant fields and hills (a poignant instance of the latter being sand-dunes), whose shapes can be understood quite accurately by the unaided eye, even though binocular vision is of little help (Figure 6.6). It is easy to check this by covering up one eye, when the sensation of depth is scarcely diminished. Similarly, a monocular view of a ball, egg or cushion can provide an accurate measure of its shape merely from the play of light on its surface: indeed, the plainer the surface, the more real and accurate is the sensation of depth. Clearly, another principle is coming into play in these instances.

To understand the process and apply it to computer analysis of 3D shapes, we must first model the reflectivities of object surfaces. For shiny objects, specular reflection provides the most useful paradigm: any light that falls on a surface is reflected in such a way that the angle of reflection equals the angle of incidence, and both rays are in the same plane as the local edge normal. For matt surfaces, the reflection is not specular, and the simplest model involves the complete absorption of the light and its re-emittance equally in all directions; in addition, the intensity of the reflected light is proportional to the cosine of the angle of incidence i. For such *Lambertian* surfaces, there is no memory of the direction of the incident light beyond that indicated by the angle of incidence, so a knowledge of the intensity only indicates the direction of the incident light within a cone around the local surface normal. Alternatively, if the direction of the incident ray is known, the intensity of the reflected light only indicates that the direction of the local surface normal must lie on the surface of a cone of semi-angle i (see Figure 6.7(a)). Thus the direction of the surface normal cannot be deduced unambiguously, and there is no direct information on the depth of the surface. In fact, in a monocular view, although it might be possible to deduce the shape of the surface, its distance or scale will remain unknown (though if either of these quantities is known the other could be computed).

Before exploring the *shape from shading* phenomenon further, it should be noted that most surfaces are neither specular reflectors nor Lambertian reflectors: most can be approximated as combinations of these, but the particular combination will probably have to be deduced from the particular images being viewed, as there is a wide variety of surface materials and also of surface conditions such as cleanness, dryness, roughness and so on. In addition, the actual surface reflectivity of a near-Lambertian surface may well be unknown, and will have to be deduced from the images in question. Space will not permit these aspects to be discussed in detail here.

Until recently it was not known how to formally analyse the reflected light patterns in order to obtain unambiguous descriptions of surface shapes. The best that could be done was to apply highly computation

a

b

Figure 6.6 Examples of 'shape from shading'. (a) Eggs whose shape is unmistakable from the shading. (b) Lentils in which the slight impression of a hand is clearly visible.

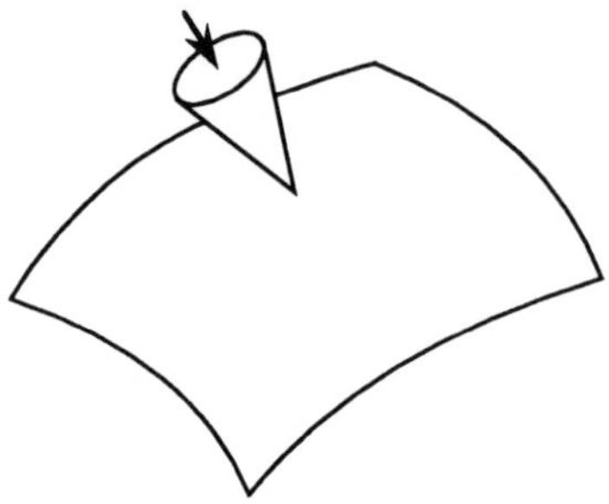

a

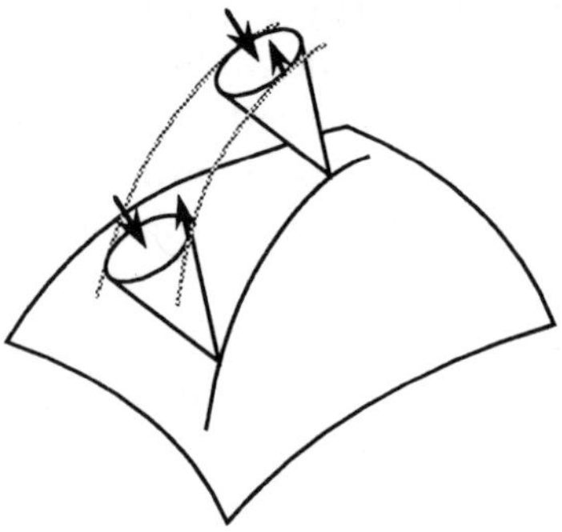

b

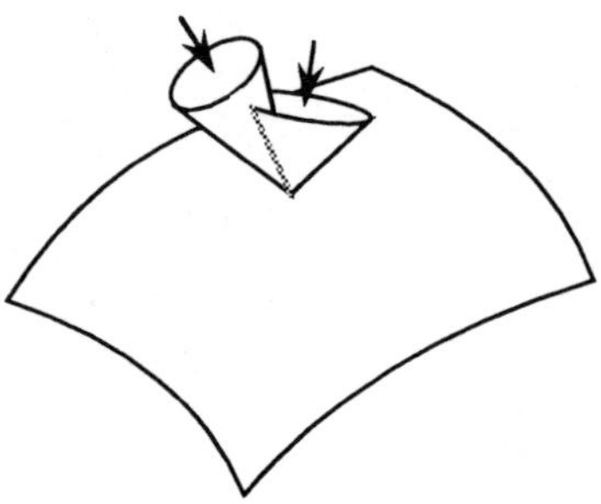

c

Figure 6.7 Computation of surface orientation. (a) For a matt surface of known surface reflectance, the apparent brightness is a function of the strength and direction of the illumination: this means that the local surface orientation vector must lie along the surface of a cone centred along the direction of illumination, represented here by the direction of the arrow. (b) When a line of constant intensity on a matt surface is viewed, the local orientations must lie on a sequence of cones similar to those shown in (a): their envelope is a pair of surfaces, one of which includes all the local orientations for points on the line of constant intensity. (c) Photometric stereo involves the sequential application of illumination from two or more directions. Here, the two cones both include the local orientation vector which is thereby clearly restricted to just two directions in space.

intensive iterative procedures which would aim to arrive at self-consistent solutions to the shape determination problem: specifically, all surface brightness values would be proportional to the cosines of the local angles of incidence, and the *two* parameters defining the orientations of the local edge normals had to be adjusted minutely until a self-consistent shape emerged which was also consistent with the observed boundary conditions; the latter concern the intensities that are possible at positions where the edge normals are perpendicular to the direction of the incident light, and where the edge normals are perpendicular to the viewing direction.

More recently, it has been noted (Cho *et al.* 1997) that the local edge normals must lie on cones, and that lines of constant brightness on a surface must correspond to sequences of cones with the same apex angle. For a smooth surface, adjacent cones must intersect at lateral positions which lead to envelopes at angles $\pm i$ to the direction of the incident light, so most of the ambiguity of the local edge normals will have been eliminated (Figure 6.7(b)). Indeed, if it is known that the surface is convex, and that there is therefore a natural direction on the surface in which the intensity is increasing, the sign uncertainty is eliminated and the direction of the edge normal is completely determined. Of course, there is the remaining problem that the absolute surface reflectivity is unlikely to be known exactly, and this may have to be estimated by some alternative method, e.g. by considering the intensities at positions where the brightness reaches a maximum. However, the procedure described above solves the basic problem.

In some situations, the shape of the surface may need to be mapped out more accurately than can be achieved by the above means. In that case a procedure called *photometric stereo* can be invoked. This involves illuminating the surface using several directional light sources in turn. After the first of these has been applied, the direction of each edge normal is known within a cone of directions, as indicated earlier. After the second has been applied, each edge normal is known within two intersecting cones, leading to one of two possible orientations in space (Figure 6.7(c)). Then after the third has been applied, the remaining ambiguity is eliminated. Furthermore, since the three cones intersect in a common line, and it can be ensured that two of them do so nearly orthogonally, the absolute direction of each edge normal in space can in principle be determined with quite high accuracy. Not only do the three intensity measurements at each point permit the two local orientation parameters to be estimated, but also they permit the surface reflectance to be calculated (Horn 1986). However, it may be useful to use a fourth light source to overdetermine the orientation solution, thereby giving the additional information needed to check how close the surface reflectance is to Lambertian.

As hinted earlier, shape from shading can at best determine local surface orientation everywhere on a surface: it still remains to compute the actual shape of the surface in terms of displacements relative to a set of world axes or a set of suitable axes within each object. However, it is important to observe that the ultimate result is a depth map of the object field, or equivalently, a space map describing the surface shape of each object in the scene. Thus we have arrived at essentially the same point that we arrived at by binocular vision or with the help of structured lighting. There are two major further steps before objects can be recognized in 3D:

the first is modelling or characterizing the shapes; the second is inferring the presence of whole objects. Comparing the situation with that for 2D, we have now progressed as far as we did in Chapter 2, where we obtained regions and boundaries delineating 2D shapes. Recognition of continuous 3D shapes such as spheres or eggs would follow much the same methods, suitably adapted, as for finding 2D shapes such as circles, albeit with the higher dimensionality introducing greater computational requirements. For reasons of space, we shall not go into this aspect further here. Instead we shall concentrate on developing the point feature-based approach of Chapter 4 so that it can be applied to 3D recognition.

6.4 Views and projections

When objects are viewed in a 3D scene, there are two particular complications that need to be tackled. One is the fact that objects appear smaller at greater distances. The other is the fact that they appear distorted by the effects of perspective. In fact, these complications are related: they arise as the rays arriving at the camera pass through a point called the *centre of projection* on their way to the image plane (off-axis rays do not actually pass through the centre of projection, but the lens still helps them to contribute to a coherent focused image by bending their paths appropriately).

Perspective is well known to produce foreshortening of objects, so that the parts nearer to the camera are magnified relative to those further away. This effect is very noticeable in such cases as close-up photographs of faces, where the nose may appear grotesquely magnified. Similarly, parallel railway lines appear to get closer and closer at increasing distances, until they become coincident on the horizon.

Perspective is hard to deal with analytically, and tends to require fairly tedious trigonometry before objects can be recognized or measured with any certainty. Fortunately, *full perspective projection* (FPP) often gives way to *weak perspective projection* (WPP) which is far easier to visualize and handle. In WPP, it is assumed that the differential depth ΔZ of any object is far less than the depth Z of the object within the scene. This means that the rays converge very little on their way past the object and can be assumed to be almost parallel at that point. In particular, if we view a cube under WPP, its back and front faces will appear to be the same size, and there will be no foreshortening. The main thing to bear in mind when using the WPP approximation is that a different WPP applies to each object in the scene, and that objects also have to be characterized by their depths within the scene: as a result, the apparent size of each object is governed by its depth in the scene. Thus a cube that is moved closer will appear larger but will have exactly the same apparent shape, unless it is moved so close that the WPP approximation breaks down (Figure 6.8).

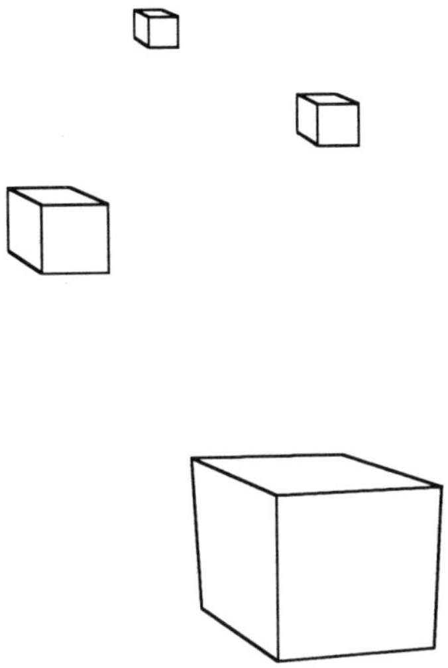

Figure 6.8 Onset of full perspective projection. In this scene containing a number of identical cubes, the front-most cube is so close that its depth in the scene Z is comparable with its differential depth ΔZ: thus the projection is no longer weak, and the object appears somewhat foreshortened.

We next examine some of the consequences of WPP when attempting to recognize objects by their point features. Let us observe three features in an image, and identify them with three features on a known object (for simplicity, we shall assume here and in what follows that all features are distinguishable—unlike the situation in Chapter 4). It turns out that the perspective distortions can be resolved by imagining the idealized flat 2D object being placed in the image, adjusted in size, and tipped through an angle α about a carefully chosen axis λ (Figure 6.9). Although this seems quite a simple situation, it is possible to see that there are always two solutions, which differ in the sign of α. The reason for this is that the adjustment for shape is achieved by a rotation α, which compresses the object by a factor $\cos\alpha$ in a direction perpendicular to the rotation axis λ: as it is $\cos\alpha$ that is specified, α itself is only defined within a sign ambiguity.

It is attractive to attempt to remove the sign ambiguity by identifying more points on the object. However, each point that is taken which is coplanar with the existing ones leads to the same equation for $\cos\alpha$, so the ambiguity is not resolved. There is a good reason for this. The position of every additional point in the plane can be deduced from the positions of the first three merely by taking a vector sum of the form:

$$\mathbf{r} = a_1\mathbf{r}_1 + a_2\mathbf{r}_2 + a_3\mathbf{r}_3 \tag{6.3}$$

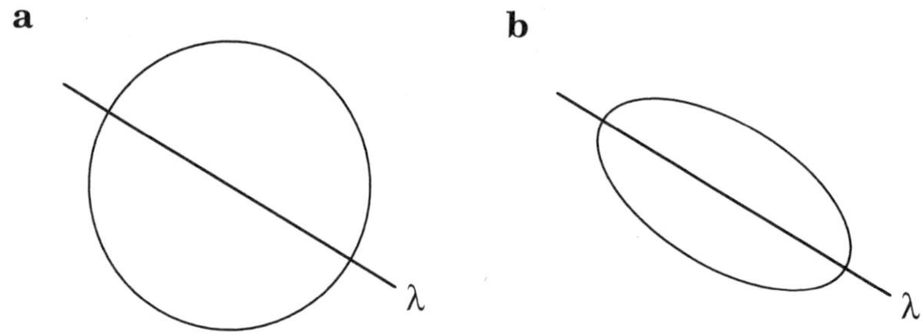

Figure 6.9 Circle viewed obliquely under weak perspective projection. Here the image of an obliquely viewed circle (b) can be imagined as formed by tipping the circle (a) through an angle α about a carefully chosen axis λ. Notice that the sign of α is not determined uniquely by the view in (b).

Once the values of a_1, a_2, a_3 are known, taking any additional point does not add any further information about the orientation of the object. There are two potential means of resolving the ambiguity: the first is to try using additional *non*-coplanar points; the second is to revert to FPP by moving the camera closer. We shall consider the latter route later. Meanwhile, the first route yields an immediate solution, as the fourth point will behave differently for the two possible α-rotations, and it will then be obvious which gives the correct solution (Figure 6.10). Clearly, once a complete unambiguous interpretation has been obtained, further points will in principle be superfluous. However, the discussion in Chapter 4 indicates that using more than four points will be beneficial in providing greater

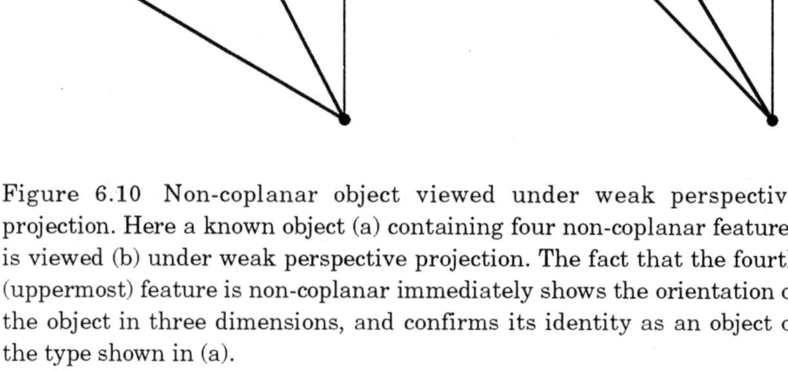

Figure 6.10 Non-coplanar object viewed under weak perspective projection. Here a known object (a) containing four non-coplanar features is viewed (b) under weak perspective projection. The fact that the fourth (uppermost) feature is non-coplanar immediately shows the orientation of the object in three dimensions, and confirms its identity as an object of the type shown in (a).

redundancy, thereby ensuring that the solution that has been found is not fortuitous. Additional points should also lead to a substantial increase in the accuracy with which the object can be located, since a greater amount of numerical averaging of the data can be achieved.

At this stage we appear to have solved all the obvious problems of object location with a mere four non-coplanar points being observed under WPP. In fact, the outstanding problem is how to identify a set of four coplanar points unambiguously. WPP has been seen to be incapable of achieving this, even if an infinite number of coplanar points are viewed. However, FPP provides a solution. To see why this is possible, let us have one more look at WPP. First, WPP provides no depth information which will help a fourth coplanar point to provide additional information. But under FPP we can test all the points to see which presents the most foreshortening; that point will then be the closest point, and it will then be clear which is the correct sign of α for the other three points.

6.5 Motion

Although 3D analysis is vitally important when dealing with real-world scenes, it is by no means the only relevant factor—another being the analysis of motion. An obvious way of studying motion is to examine the differences between the successive frames obtained by a TV camera. While this is an obvious approach, there are certain difficulties to be overcome.

First, consider a plain rectangular object moving parallel to its two longest sides (Figure 6.11). These two sides will actually give no indication of motion as the various points on them are indistinguishable and will give zero output in the difference image. In addition, the various parts within the object give no indication of motion for the same reason. It is only the shortest sides of the object which are respectively covering up and uncovering parts of the background that give any indication of motion. Next, examining local parts of these regions shows that motion is occurring, but in fact only permits the component of the motion parallel to the longest sides to be estimated: any component parallel to the shortest sides will be undetectable by making local measurements on them (this is known as the *aperture problem* and arises because only restricted information is available within a small aperture—Figure 6.11(e)). Finally, motion at any locations on the shortest sides which lie adjacent to background regions of the same intensity will not be detectable. Thus there are quite strong restrictions on what aspects of the motion can be measured by image differencing.

We can analyse the situation by considering the intensity function $I(x,y,t)$ and expanding it in a Taylor series:

$$I(x+dx, y+dy, t+dt) = I(x,y,t) + I_x dx + I_y dy + I_t dt + \ldots \tag{6.4}$$

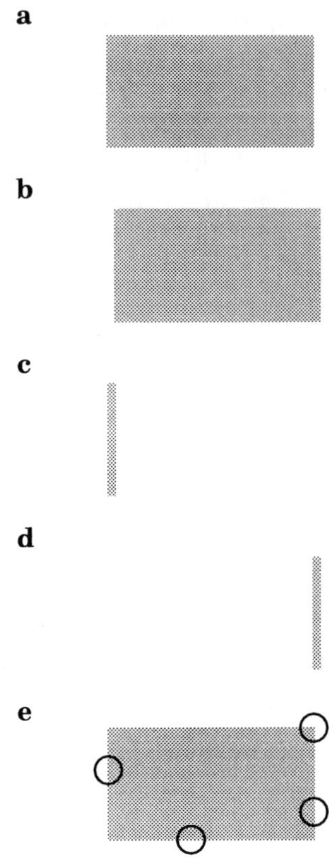

Figure 6.11 Motion and the aperture problem. (a) shows a moving rectangular object which appears shifted in (b). (c) and (d) show negative and positive difference images, corresponding respectively to parts of the original object that are removed and added between the two frames. (e) shows a set of circular apertures which are viewed locally to deduce the motion: the aperture at the bottom edge (which is parallel to the direction of motion) gives no indication of motion, while the two in the middle of the left and right edges give no information on possible components of motion along the directions of these edges. Only the aperture over the top right hand corner would be able to provide enough information to estimate the exact motion vector.

where second and higher order terms will be ignored. In this equation I_x, I_y, I_t denote respective partial derivatives with respect to x,y and t. Next, imagine that the image has shifted locally by amount (dx,dy) in time dt, so that it is functionally identical at $(x+dx,y+dy,t+dt)$ and (x,y,t):

$$I(x+dx,y+dy,t+dt)=I(x,y,t) \tag{6.5}$$

We now deduce:

$$I_t = -(I_x \dot{x} + I_y \dot{y}) \tag{6.6}$$

This equation can be written more succinctly in the form:

$$I_t = -\nabla I . \mathbf{v} \tag{6.7}$$

where **v** is the local velocity. We can sum up the situation by noting that motion is only measurable at locations where there is a strong intensity gradient *coupled with* a strong vector component of the motion parallel to the intensity gradient. When an object is plain, i.e. it has a uniform intensity over its surface, the only positions where the intensity gradient is non-vanishing are at its edges, so it is only at locations where the edge normal has a component parallel to the direction of motion that permit the motion to be estimated.

In principle, knowledge of $I_t(x, y)$ should permit the velocity field $\mathbf{v}(x, y)$—or the *optical flow field* as it is known—to be estimated. However, as noted earlier, the aperture problem only permits one of its local components to be calculated. Ultimately, the reason for this is that it is impossible to calculate a 2-component vector from a single scalar equation (equation (6.7)). An accepted solution to the problem is to use an iterative approach (viz. relaxation labelling) to arrive at a consistent global solution which minimizes global error. In some circumstances the problem is ill-conditioned and the solutions are inaccurate and are arrived at slowly: however, in a textured image this difficulty should not occur.

Once an optical flow field has been obtained, interpretation is not necessarily a trivial matter, as a number of objects including the camera itself may be moving. If the camera is moving forward and all the objects in the scene are stationary, they appear to diverge from a single point in the image: this point is called the *focus of expansion* (Figure 6.12). In fact, this situation is indistinguishable from an image seen by a stationary camera when all the visible objects are moving towards it with equal velocities (as might happen in space during a meteor storm).

Another interesting case arises when a large object is moving towards the camera: assuming it is a rigid object, all parts of it will have the same velocity, though parts of it will be at different depths in the scene and will therefore present different motions from those which would be evident for features at the same depth near the front of the object. In fact, features which are closer to the camera will appear to have faster lateral motions. However, features which are further from the focus of expansion will appear to move more rapidly even if they are at the same depth in the scene. This gives the possibility of estimating object structure from optical flow (Figure 6.13). We here quote the relevant formula, derivation of which may be found in more advanced texts (e.g. Davies 1997e):

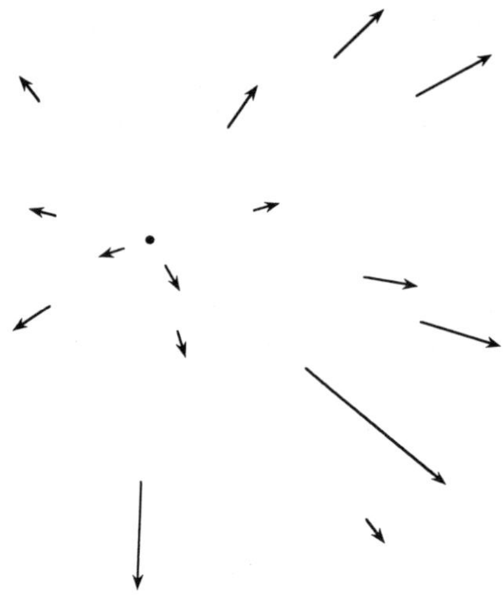

Figure 6.12 Focus of expansion. This diagram shows the apparent motion vectors in the image for objects approaching the camera. The focus of expansion is the black dot showing the position in the image from which all the motion vectors appear to originate.

$$\frac{Z_1}{Z_2} = \frac{r_1/r_2}{\dot{r}_1/\dot{r}_2} \tag{6.8}$$

where r_1 and r_2 are the image distances of features 1 and 2 from the focus of expansion.

What is important about the above formula is that it permits the apparent motion parameters to be used to determine the relative depths of

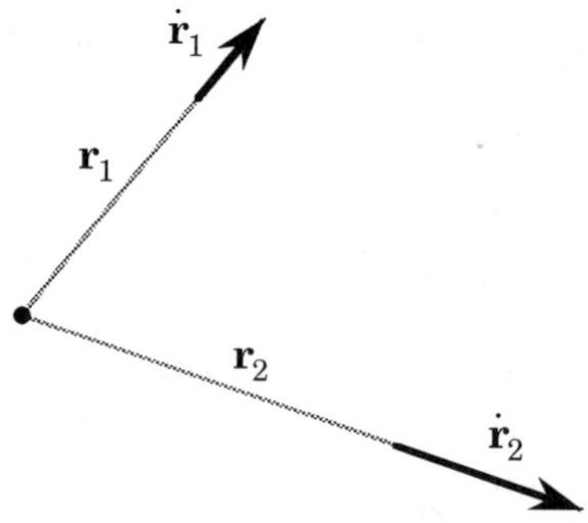

Figure 6.13 Geometry for computation of depth from motion. Here two object positions and their motion vectors relative to the common focus of expansion, indicated by the black dot, are shown. The magnitudes of the four vectors provide sufficient information to permit the relative depths of the two objects to be estimated (see text).

object features. Notice that only the relative depths can be obtained in this way because a single image does not permit absolute depth to be determined, though if one dimensional measure of the object is known (e.g. its frontal width), all other measures could be calculated.

Before leaving the topic of optical flow, note that we have considered only simple cases of relative motion in which bodies have fixed velocities and are not rotating. In general, rotations add considerable further complexity to the problem, as the number of interlinked parameters required to explain the complete motion of the observed system can be quite high. However, in the special case when all linear velocities are zero but an object is rotating either about the line of sight or about an axis perpendicular to it, it is possible to interpret the scene without especial difficulty. The reader is referred to more specialized texts for fuller explanations of more general cases.

The subject of optical flow has arisen because it is straightforwardly calculated by local operators, though some complexity arises because of the need to obtain both components of the velocity vector at each point in the image. However, the incidence of low intensity gradients in some parts of an image means that optical flow is poorly estimated in these places and cannot be relied upon for image interpretation. In fact, we have already seen that this is a common situation in image processing, as ordinary (non-motion) edges often peter out, and reliance then has to be placed on other parts of the image for interpretation and recognition functionality. Thus there has to be significant reliance on inference: in this light the lack of motion data around parts of an object boundary will be of little consequence if its velocity can be hypothesized from other boundary locations such as corners (Figure 6.11(e)). Space does not permit us to discuss these points in more detail here.

6.6 Concluding remarks

It is a common idea that binocular vision solves all the problems of 3D perception. It has been one of the aims of this chapter to show that binocular vision is not a panacea, and to indicate some of the main difficulties of 3D vision—not least the correspondence principle on which binocular vision depends. In fact, 3D vision is a vast subject which warrants a book in itself, and we have had to be content with a summary of just a few pages. The reader may well benefit by referring to Davies (1997e) for a more detailed but still readable account in the space of four chapters.

In the food industry, structured lighting has been used quite widely, e.g. for analysis of the 3D shapes of loaves of bread (Chapter 9), and in this context the early book by Shirai (1987) is still a useful reference. Horn's (1986) book provides an in-depth study of the shape from shading problem, though notable advances (e.g. Cho *et al.* 1997) have been made since it was written.

Over the past few years the *perspective N-point problem* (finding the pose of objects from N features under various forms of perspective) has been studied in some depth. Fischler and Bolles described the key RANSAC algorithm in 1981, though they did not discuss pose determination under WPP. The latter has been investigated by Horaud *et al.* (1989): see Davies (1997e) for an overview of the FPP and WPP matching problems.

Another subject which is highly relevant to tackling the complexities of 3D vision is that of invariance under perspective projection. Although a key volume of edited works is available on this topic (Mundy and Zisserman 1992), this is now dated and difficult to understand for those outside the immediate area. Unfortunately, invariance is not a topic which could be broached here, for lack of space; this should not be taken to imply lack of utility, but rather that the mathematical basis of the subject takes at least 10 pages to set up and is not embarked upon lightly: see Davies (1997e) for a readable summary.

Motion is another topic which requires a certain amount of mathematical preparation, and it was felt useful to include something on it in this monograph, in order to help prepare the reader for future events in the food industry—some of which are already apparent in Chapter 12. For further details of optical flow and related topics, the reader is referred to the wider literature (e.g. Nagel 1983, 1986; Hildreth 1984; Brady and Wang 1992; Faugeras 1993). Other notable work in this area includes the application of snakes to tracking (Delagnes *et al.* 1995), the use of Kalman filters in tracking (Marslin *et al.* 1991), and the recognition of vehicles on the ground plane (Tan *et al.* 1994).

Chapter 7
Pattern recognition

7.1 Introduction

The early chapters in this monograph may have left the impression that object recognition can normally be achieved by performing a sequence of image processing operations which end naturally with circle centres being located, boundary shapes being categorized or the presence of objects being inferred from their feature locations. There has been little room for ambiguity of interpretation, except when partial overlap or occlusion takes place. Indeed, it would appear that the process of systematically removing redundant information from images eventually leaves only the data that is required for recognition, and that recognition is a totally obvious and automatic process. For example, image thresholding followed by thinning and skeleton analysis follow this line of removing enormous amounts of redundant information, leaving just a few nodes on each shape which would ideally be sufficient to mediate recognition. However, this view does not reflect the true situation. First, object shapes tend to be more variegated than this viewpoint suggests: for example, when examining printed As, there are upper and lower case versions, large numbers of fonts, with and without serifs (which are an obvious source of additional skeleton junctions), and italicized variants, not to mention the different possible font sizes: note also that even the lower case A has two common versions, with or without a branch over the top. And of course there is an infinity of hand-written versions, including an infinity of potential variants from the same hand. Second, even if the same paradigm versions are adhered to, the effect of noise can be so serious as to cause major differences between one sample and the next. One such difference that can arise with the skeleton based approach to recognition is that a single hole caused by noise can radically alter the shape of the skeleton and the number of nodes it contains: such distortions can quickly make recognition well nigh impossible via this route.

To overcome these problems we have to give up the idea that recognition is a purely deductive process: even considering it an inferential process is likely to be a poor strategy, as this will only yield a list of possible interpretations, whereas what we want is the most *probable* classification in each case. Having arrived at this position, it seems attractive to go further, and regard object shapes as bit-maps, and to consider each bit-map as having a class—or rather a list of possible classes, each with a known probability—which may be looked up in a suitable dictionary.

There are three major problems with this idea: the first is how to combine the resulting classifications and probability tags to arrive at the most probable interpretation of the image as a whole (whether the image consists solely of a set of characters or a natural outdoor scene); the second is how to obtain a viable set of look-up tables; and the third is how to cope with the fact that the look-up tables will be very large indeed in many practical applications. In fact, the last two of these problems will often be insuperable, as the number of binary patterns of size $N \times N$ is 2^{N^2}, which is cripplingly large for N as small as seven. Thus a totally different line of attack must be used. First, it pays to carry out as much structural pattern recognition as possible before resorting to probabilistic or statistical pattern recognition (SPR). For example, even the simple procedure of eliminating noise and normalizing the positions of characters within a subimage should ease the SPR task and bring it within the realm of the possible. Next, SPR can be split into two parts—feature measurement and final classification: treating the original bits or pixel intensities as features is unlikely to be the best strategy; it is normally far better to determine the lengths, widths, aspect ratios, moments, mean intensities, or numbers of skeleton nodes of various types, and to use these parameters as data to be fed to the final classifier. How to design such a system has not been part of the previous discussion and will be considered in the following sections.

7.2 Bayes' approach to SPR

The process of SPR can best be illustrated by considering a two-class system with a small number of features. We need not worry at this stage what the features are: they can be taken to include quantities such as length, width, intensity and a colour parameter. (In fact, there is no need for them to be continuous variables: digitized or bit variables are also permissible, though here it will be easiest to imagine them as continuous variables.) In the two-class case, incoming patterns may be marked with a single feature x, with class 1 patterns having low values of x and class 2 patterns having higher values of x. If it is possible to achieve error-free separation by applying a single threshold, the situation will be deterministic and there will be no SPR problem. However, if the distributions present some overlap, as in Figure 7.1(a), it will not immediately be clear how to achieve an optimal solution. One idea is to eliminate as much error as possible by classifying patterns as class 1 on the

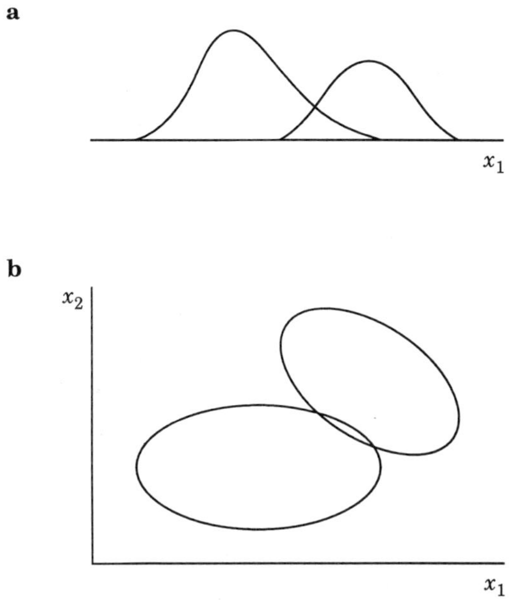

Figure 7.1 Overlap of class distributions. (a) shows the overlap that might be expected between two class distributions if a single feature x_1 is used; (b) shows the situation when two features x_1, x_2 are employed. It is clear that (b) offers increased opportunity for minimizing classification errors by appropriate choice of class boundaries.

left of the crossover point on the histogram, and as class 2 to the right of it. Much depends on how the histogram distributions were constructed and what they actually represent. An inkling of a deeper problem is that if the patterns originate from a set of ripe tomatoes intermixed with a much smaller number of slightly less ripe ones, then it may be numerically more accurate to guess that they are ripe than to risk quite large errors in the measurement of ripeness from the colour. Clearly, we should take account not only of the colour but also of the much higher *a priori* probability that the tomatoes will be ripe, as this should give a more accurate overall assessment. Bayes' rule permits this to be achieved via a single equation:

$$P(C_i|x) = p(x|C_i)P(C_i)/p(x) \tag{7.1}$$

where

$$p(x) = \sum_j p(x|C_j)P(C_j) \tag{7.2}$$

Here, $P(C_i)$ is the *a priori* probability of class C_i, $p(x)$ is the probability density for feature x; $p(x|C_i)$ is the class-conditional probability density for feature x in class C_i, i.e. the probability that feature x arises for an object known to be in class C_i; and $P(C_i|x)$ is the *a posteriori* probability of class C_i when x is observed.

In any application, $P(C_i|x)$ is the quantity we wish to know: to calculate it we need to know both the intrinsic probability $P(C_i)$ of C_i and the distribution $p(x|C_i)$ of feature values that arise for this class. These probabilities have to be obtained by prior measurements on the *training set*, which has to be a fair representation of the objects of the various classes. Clearly, such a classification scheme relies on the quality of the training set, and great care must be taken in compiling it to ensure that it is representative of all patterns that could arise during testing.

Nevertheless, a Bayes classifier based on a single feature will in general not lead to perfect classification, and it is fairly obvious that a reduction in classification errors will be possible if we take two or more features, as shown in Figure 7.1(b). Perhaps surprisingly, taking an overly large number of features can result in a deterioration in performance, since in the end the additional features will not be statistically significant (i.e. they will not be specified sufficiently accurately by the available training set and will only serve to introduce uncertainty).

A more general form of Bayes' rule is required to cover the case where there are several features in a multidimensional feature-space, by using the modified formula:

$$P(C_i|\mathbf{x}) = p(\mathbf{x}|C_i)P(C_i)/p(\mathbf{x}) \tag{7.3}$$

where $\mathbf{x}$ represents the multidimensional feature vector, and $p(\mathbf{x})$ is the overall probability density:

$$p(\mathbf{x}) = \sum_j p(\mathbf{x}|C_j)P(C_j) \tag{7.4}$$

Once the values of all the $P(C_i|\mathbf{x})$ have been computed, we can determine the class corresponding to $\mathbf{x}$ as C_i if:

$$P(C_i|\mathbf{x}) > P(C_j|\mathbf{x}) \text{ for all } j \neq i \tag{7.5}$$

In fact, this maximum probability classification can be shown to lead to a minimum error classification rule. However, sometimes a minimum cost classification rule is more appropriate, where the cost is some function of the errors involved in making correct or incorrect decisions. Thus cost will be associated with a loss $L(C_i|C_j)$ corresponding to making a judgement C_i when the true class is C_j. The overall problem is then to minimize the overall loss by determining the value of i for which the following expression is lowest:

$$\mathscr{L}(C_i|\mathbf{x}) = \sum_j L(C_i|C_j)P(C_j|\mathbf{x}) \tag{7.6}$$

Typically, $L(C_i|C_j)$ will be zero if $i = j$, and some fixed substantial value (which we could notionally take to be unity) if $i \neq j$. Since these considerations will be highly dependent on the particular application, we shall not take the matter further here.

Next, we note that Bayes' approach can be used to take account of context. This is possible because context affects the *a priori* probability for the object currently being considered. Perhaps the simplest example of this is in OCR, where the probability of a given character appearing depends on the digram, trigram and higher relative frequencies for the specific language.

Finally, Bayes theory leads naturally to discriminant analysis (DA) and multivariate analysis (MVA), the latter being a generalization of DA to the multiple parameter case. In DA (and MVA) the training data is modelled, e.g. in the form of Gaussian distributions, and discriminant functions (DFs) are devised by which to decide the class of any data point on the basis of the measured parameter(s). Much is known about how to select suitable DFs to optimise classification accuracy (see, for example, Duda and Hart 1973). Section 12.5.2 gives an example of this type of modelling, and Chapters 9, 10, 12, 13 describe applications of DA and MVA.

7.3 The nearest neighbour approach

While the Bayes' approach to SPR arises naturally when distributions of feature values are considered, an alternative approach arises from consideration of the individual training patterns: this is the *nearest neighbour* (NN) algorithm. It involves first storing all the training set patterns as feature vectors in a multi-dimensional feature space, and tagging them with their classes; and then testing any new pattern by determining which training pattern it is closest to in the feature space, and assigning it the class label of this pattern (Figure 7.2(a)).

This simple procedure is straightforward to imagine and to program, and it turns out to have an error rate within a factor two of that of the Bayes' classifier. Thus it essentially possesses the main advantage of the Bayes' classifier, though careful calculation of probability density functions is not required. However, it is no surprise that much attention focused on the NN algorithm at an early stage in the development of SPR.

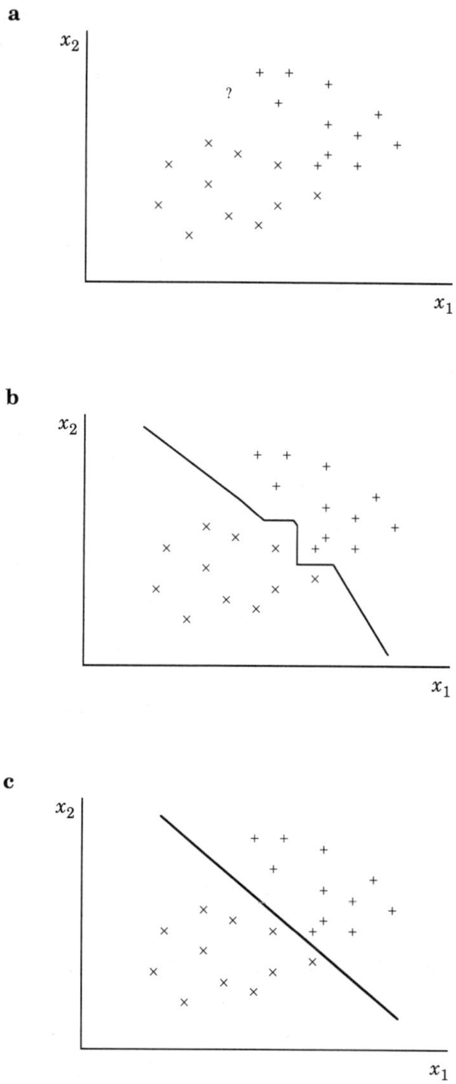

Figure 7.2 Nearest neighbour and related classifiers. (a) demonstrates the operation of the nearest neighbour classifier. Two regions are represented by sets of patterns forming a training set, these being marked with plusses and crosses in the x_1, x_2 feature space; the question mark indicates the position of a test pattern, and will be assigned to the *plus* class as its closest neighbour in feature space is a plus. (b) shows that exactly the same classification performance can be achieved by a suitable *piecewise linear* classifier. (c) shows how the piecewise linear classifier of (b) can be approximated by a linear classifier, with no apparent increase in error rate. However, in general, the error rate will increase for (c). Note also, that the data presented in (a) is separable, though this will not generally be the case in practice (see Figure 7.1).

Furthermore, it was soon found how to improve the error rate to match the Bayes' error rate as closely as could be required by generalizing the NN algorithm to take in the *k* nearest neighbours in feature space (the class label of the test pattern was then made equal to the majority class for its *k* nearest neighbours).

To underline the operation of the NN algorithm, note that when the class distributions overlap in feature space, the training set patterns will be interwoven, though if the density of training set patterns is high enough, error rates from this source can still be close to those for the Bayes' approach.

Unfortunately, the NN algorithm, and *a fortiori* its derivative, the *k*-NN algorithm, requires considerable storage and a correspondingly large amount of computation, since in principle every test pattern has to be compared with every training set pattern that is stored in feature space. Methods have been found for ameliorating these problems, e.g. by pruning the training set (which involves the elimination of patterns which are not near to class boundaries in feature space).

A different and rather general alternative strategy to pruning the training set patterns is to define and store information on the inter-class boundary positions. In a linear classifier these are stored as linear hyperplanes in a multi-dimensional feature space (Figure 7.2(c)). This requires a minimum of data per hyperplane, thus eliminating the need for storage of a great many training set patterns, though the approach necessarily increases error rates. The reason for this may be seen from Figure 7.2(b), where the training set patterns close to class boundaries actually constitute a piecewise linear classifier, with (in general) one hyperplane segment per pair of boundary training set patterns. Thus an important aspect of producing a linear classifier is that of finding an optimum balance between storage and error rate. The other important factor is that of cutting down computational load, but clearly this is reduced more or less in proportion to the storage: we shall not dwell on details of the situation further here. However, it is worth remarking that the piecewise linear classifier is not the only possibility: any suitable functional classifier that matches the multi-dimensional data boundaries can be chosen.

In the above discussion the impression may have been conveyed that piecewise linear classifiers are developed by smoothing out the decision surfaces obtained from a NN classifier. Although this is a possible approach, it will generally be far too computation intensive, and the method adopted in practice is to start with a system of planar class boundaries which are made to move as successive training set patterns are presented to a suitable training algorithm. The design of such algorithms is by no means trivial, not least because they have to converge systematically even though successive training set patterns may provide conflicting information—as happens when patterns from one class appear within regions populated mainly by another class. In the following section we shall see how such problems can be resolved.

7.4 Artificial neural networks

The concept of an artificial neural network (ANN) has spurred on much research in the area of SPR. It is based on the idea that the neurons in the human brain perform complex cognitive tasks such as recognition, and achieve this by training rather than by overt programming—the likely simplicity of each neuron naturally being offset by the prodigious numbers of neurons in the brain and by the impressive connectivity patterns between them (it has been estimated that the brain possesses well over 10^{10} neurons, each having some 10^4 connections to other neurons). One of the first models for an artificial neuron was Rosenblatt's *perceptron* (1958, 1962), which has had enormous influence on the subject.

The *simple perceptron* is a linear classifier that classifies patterns into two classes (Figure 7.3(a)). Its input is a feature vector $\mathbf{x} = (x_1, x_2, \ldots, x_N)$, and it produces an output $\sum_{i=1}^{N} w_i x_i$ to which a threshold θ is applied. The mathematics is simplified by writing $-\theta$ as w_0, and taking it to correspond to an input x_0 which is set at unity. The linear output thus takes the form:

$$d = \sum_{i=1}^{N} w_i x_i - \theta = \sum_{i=0}^{N} w_i x_i \tag{7.7}$$

and the final decision is given by applying a step-function f:

$$y = f(d) = f\left(\sum_{i=0}^{N} w_i x_i \right) \tag{7.8}$$

One way of training this type of neuron is to use the *fixed increment rule*, which aims to improve the error rate by repeatedly moving the linear discriminant plane a fixed distance towards a zero misclassification position. However, this type of approach is only successful if the features are linearly separable, and can therefore only be used in a minority of practical applications.

The Widrow-Hoff delta rule overcomes this problem by making changes in the weights proportional to the classifier error $\delta = C - d$, C being the actual class and d being the value predicted by equation 7.7:

$$w_i(k+1) = w_i(k) + \eta \delta x_i(k) = w_i(k) + \eta [C(k) - d(k)] x_i(k) \tag{7.9}$$

In this way the Widrow-Hoff rule makes no assumptions about the separability or otherwise of the data: it makes a judgement about the best position to place a decision surface by weighting all the data appropriately. In addition, it does not become 'satisfied' when separation has been achieved, but continues to move the decision surface to where it estimates the minimum error position is situated: thus it is sensitive to the underlying class probability distributions.

To cope with many classes of pattern, it is possible to use a *single layer perceptron* consisting of many single perceptrons, each capable of dichotomizing the feature data between pairs of classes (Figure 7.3(b)).

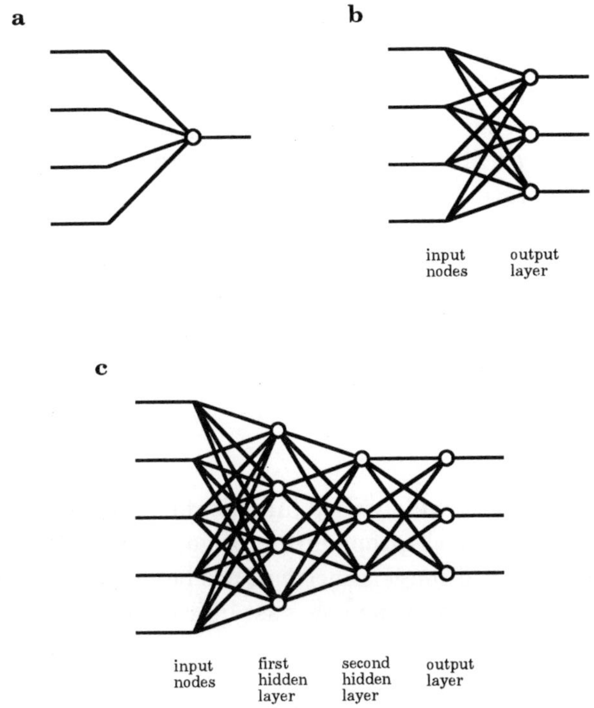

Figure 7.3 Feedforward neural networks. (a) shows a *simple perceptron*, which forms the basis for many types of neural classifier. (b) shows a *single layer perceptron*, constructed from a single layer of simple perceptrons; when appropriately trained, this is able to distinguish between a number of types of pattern, typically giving a high response in the appropriate output line. (c) shows a *multi-layer perceptron*, which in principle offers improved classification behaviour when pattern classes are bounded by complex regions in feature space. In (b) and (c) the input 'nodes' are mere connection boxes: only the true neuron computational elements (containing trainable memory) are denoted by circles in these diagrams.

However, this strategy would use far too many simple perceptrons: a better strategy is to employ a smaller number of simple perceptrons in a *multi-layer perceptron* (MLP), as shown in Figure 7.3(c). Unfortunately, though Rosenblatt proposed such networks in the 1960s, it was not until 1986 that Rumelhart *et al.* were able to suggest a systematic approach, now well-known as the back-propagation algorithm, for training MLPs.

The difficulty in training a MLP lies in the *credit assignment problem*, which prevents it from being known which neuron weights can be relied upon, and which are erroneous, while training is proceeding. In particular, a neuron near the centre of the network can get misleading inputs from the initial layers of the network; worse still, it is not at all clear what its target output values should be, since by the time they reach the output of the network they have passed through later layers which distort them in possibly incorrect ways: thus there is no obvious way of assigning credit to any neuron for either correct or erroneous judgements. The way this problem was overcome was by ensuring that the overall network output depended slightly on all the neurons in the network: they were prevented from giving absolute judgements but instead permitted only to give functionally smooth variations of their inputs. In this way partial differential equations could be set up to optimize the whole network. To achieve all this, the threshold activation function had to be changed from a step function to a smoothly varying function, and the sigmoid function:

$$s(u) = \frac{1}{1+e^{-u}} \tag{7.10}$$

rapidly became the most widely used function for this purpose (Figure 7.4).

Details of the back-propagation algorithm and various means commonly used to speed convergence will not be discussed here: the reader is referred to other texts and the now vast literature of the subject for these

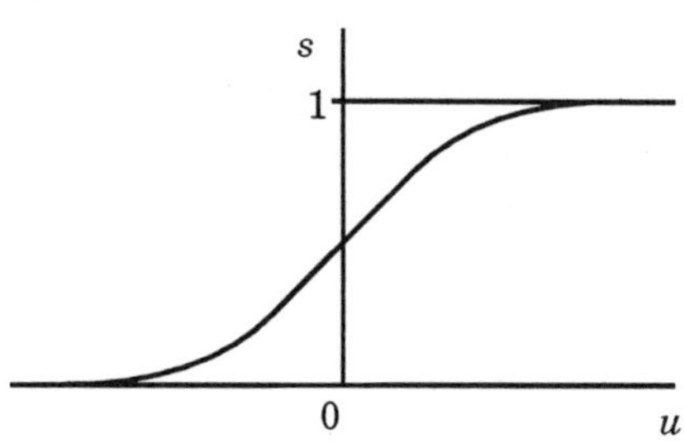

Figure 7.4 Sigmoidal activation function. This diagram shows the sigmoidal *activation function* $s(u)$ typified by the function of equation (7.10).

aspects (see Section 7.7). However, it will be worth pointing out one or two facts about MLP architectures. First, Lippmann (1987) showed that a 2-layer (single hidden layer) MLP can implement arbitrary convex decision boundaries, and indicated that a 3-layer (2-hidden layer) network is required to implement more complex decision boundaries. Later, Cybenko (1988) found that it should never be necessary to use more than two hidden layers, as a 3-layer network containing sufficient neurons can cope with quite general situations. This sort of information is very useful, as the back-propagation algorithm suffers from long convergence times and significant error when several layers of neurons have to be trained 'through' each other, so any reductions that can be achieved in the numbers of layers will be valuable.

Finally, it should be noted that there is significant possibility of over-adapting ANNs to the training data. This is intrinsically the same problem of using too many features which was mentioned in Section 7.2, and applies to any SPR system, however constructed. To prevent this from happening, it is better not to train the ANN for too long on the same data-set, but instead to train if necessary on additional data. To test for over-adaptation, it is useful to hold back some of the available data to form a *validation set*, which will indicate whether ANN performance is indeed still improving during training or whether capability for generalization is starting to deteriorate. Again, the reader is referred to other texts for further details of such problems (see Section 7.7).

7.5 Supervised and unsupervised learning

In this chapter it has so far been assumed that each feature vector must be accompanied with a class tag while training a classifier. However, there are two types of training procedure—*supervised learning* in which this is carried out, and *unsupervised learning* in which it is not attempted. The main reason for adopting the unsupervised learning strategy is to avoid tedium and error for human operators who have to label each object that is presented to the classifier during training. This circumstance is quite common in industrial inspection applications, for example. If we imagine that products are flooding along a conveyor, and that a classifier is being trained to find faulty products, it would be of some advantage if the classifier learnt for itself what types of products are arriving and what natural groups they fall into. Clearly, the classifier cannot on its own know which are the faulty groups (unless all minority groups are taken to be faulty), but if a human operator were eventually presented with samples from each of the groups, he could label them appropriately, with minimum effort and hopefully with negligible error. An advantage of this approach is that the machine could find out characteristics of the data that the human operator may not know in advance: again, in an inspection application, it may well not be known quite what faults could occur on the line, and therefore what features should be looked out for in the input data.

The main problem with unsupervised learning is to achieve a reliable grouping of the products. This grouping, which is carried out in feature space, is called *clustering*, and relies on clusters of training set patterns being grouped together if they are related in some way (e.g. by high quality), and being well separated if they belong to other clusters which have rather different properties (such as having prominent defects of one sort or another). Thus the main problem of clustering is to separate training set patterns into meaningful clusters which represent statistically different properties.

Forgy's (1965) method will be used to illustrate the process of clustering. First, a number of cluster centres are decided upon using *a priori* considerations. Then each training set pattern is assigned to the closest cluster centre and the positions of the latter are recalculated. If significant changes have occurred during this process, it can be repeated as many times as necessary. It will be clear that the method has certain problems: in particular, the final result will depend on the order of presentation of the training set patterns; it will also depend on the number of cluster centres initially decided upon and on their exact locations in feature space. The ISODATA algorithm (Ball and Hall 1966) is similar to Forgy's method but is able to merge clusters which are close together, and to split elongated clusters, thereby enhancing performance.

Another type of clustering algorithm is the non-iterative type which aims to save computation by using a minimum number of passes over the training set. MacQueen's *k-means* algorithm (MacQueen 1967) involves just two passes over the training set patterns, the first pass being required to find the cluster centres and the second pass to perform the final classification. However, it is generally acknowledged that non-iterative clustering algorithms are more sensitive than iterative algorithms to the order in which data are presented. To some extent this problem can be ameliorated by randomizing the order of presentation: this is especially relevant with image data, where large tracts of any image or sequence of images will be devoted to plain background (sky, conveyor, etc.) or to standard types of object.

It will be clear from the discussion so far that algorithms such as Forgy's or MacQueen's algorithm have no intrinsic means of determining k. However, if they are tested in several phases with different values of k, the value giving the best performance according to some pre-specified criterion can be maintained and optimized, e.g. by further training on a fuller data set (see for example Zhang and Modestino 1990).

7.6 Principal components analysis

Ultimately, unsupervised learning involves a study of the nature of the data in the training set. Cluster analysis is one way of studying the data: principal components analysis (PCA) takes the process further, by determining the principal axes of any cluster and the variance of the data along the various axes directions. First the mean of a cluster is found; then a set of mutually perpendicular axes through the mean is identified; the first of these axes must maximize the variance when the data is projected onto it (Figure 7.5); subsequent axes are defined in turn, so that each in its turn has maximum variance when the data is projected onto it. At the end of the process a complete set of N principal axes is known, each with an associated variance smaller than the previous one. The set of orthogonal principal axes is the one which diagonalizes the covariance matrix $\mathbf{R}$ of the input data:

$$\mathbf{R} = \left\langle (\mathbf{x}_{(k)} - \mathbf{m})(\mathbf{x}_{(k)} - \mathbf{m})^{\mathrm{T}} \right\rangle \tag{7.11}$$

Here, $\mathbf{x}_{(k)}$ is the location of the kth data point, and $\mathbf{m}$ is the mean of the K data points; $\langle\cdot\rangle$ indicates the expectation value for the underlying population. We can estimate $\mathbf{R}$ from the equations:

$$\mathbf{R} = \frac{1}{K}\sum_{k=1}^{K} \mathbf{x}_{(k)}\mathbf{x}_{(k)}{}^{\mathrm{T}} - \mathbf{m}\mathbf{m}^{\mathrm{T}} \tag{7.12}$$

$$\mathbf{m} = \frac{1}{K}\sum_{k=1}^{K} \mathbf{x}_{(k)} \tag{7.13}$$

In fact, $\mathbf{R}$ must be real and symmetric, and it is possible to diagonalize it using an orthogonal transformation matrix $\mathbf{A}$, obtaining a set of N orthonormal eigenvectors $\mathbf{u}_i$ with real, positive eigenvalues λ_i given by:

$$\mathbf{R}\mathbf{u}_i = \lambda_i \mathbf{u}_i \qquad (i = 1,2,\ldots,N) \tag{7.14}$$

The vectors $\mathbf{u}_i$ can be obtained from the original vectors $\mathbf{x}_i$ by:

$$\mathbf{u}_i = \mathbf{A}(\mathbf{x}_i - \mathbf{m}) \tag{7.15}$$

The inverse transformation needed to recover the original data vectors is:

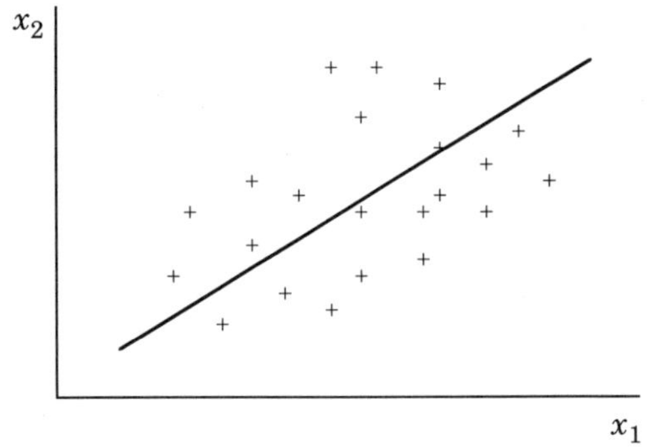

Figure 7.5 Basis of principal components analysis. This diagram shows a cluster of training set patterns in a 2D feature space. Also indicated is the line, passing through the centroid of the training set patterns, which maximizes the variance when the data is projected onto it.

$$\mathbf{x}_i = \mathbf{m} + \mathbf{A}^{\mathrm{T}}\mathbf{u}_i \tag{7.16}$$

In fact it may be shown that **A** is the matrix whose rows are formed from the eigenvectors of **R**, and that the diagonalized covariance matrix **R**′ is given by:

$$\mathbf{R}' = \mathbf{ARA}^{\mathrm{T}} = \begin{bmatrix} \lambda_1 & 0 & \cdot & \cdot & \cdot & 0 \\ 0 & \lambda_2 & & & & 0 \\ \cdot & & \cdot & & & \cdot \\ \cdot & & & \cdot & & \cdot \\ \cdot & & & & \cdot & \cdot \\ 0 & 0 & \cdot & \cdot & \cdot & \lambda_N \end{bmatrix} \tag{7.17}$$

In an orthogonal transformation, the trace of a matrix is invariant, so the trace of the input data is:

$$\text{trace}\,\mathbf{R} = \sum_{i=0}^{N} \lambda_i \tag{7.18}$$

where the λ_i are the variances of the data in the directions of the principal component axes.

In PCA, it is usual for the eigenvalues λ_i to be placed in a reducing ordered sequence, so that λ_1 represents the most significant variation in the data set, and λ_N represents the least significant variation. In many cases λ_N would be close to zero and would have negligible significance, essentially corresponding to noise. There is now the possibility of ignoring all the principal components that have negligible significance, retaining only those which are clearly important for a description of the data, and saving storage and computation in the subsequent analysis. In fact, the purpose of PCA is largely to reduce the dimensionality of the problem from N to a smaller more convenient value M, and also to provide a useful degree of data compression, or a reduction in the redundancy of the input data.

PCA has found wide application in the analysis of multi-spectral images, where it has typically been used to extract the three most important information carrying images from a set of six principal components, these being derived from original images in a set of six spectral bands. This approach has been used to highlight regions such as fields, roads and rivers which are useful for map-making purposes, to name one example. PCA has also been of some value in the analysis of textures, using the Ade approach (see Chapter 5). It has recently also been applied to the analysis of shapes of faces and animals, with considerable success (see Chapter 12). In fact, the method has proved itself very versatile, in enabling the most apparently significant sources of variation in the input data to be identified. However, some caution is needed, as it is a slightly *ad hoc* procedure, and there is no obvious reason why it should automatically lead to the most significant variations in all application areas. In particular, in cases where the data displays significant degrees of non-linearity, or where clusters can appear in close proximity, or even in contact with each other, it would be surprising if it always gave optimal results. We summarize the situation by noting that PCA is a method of representing data which is not necessarily appropriate in all cases for pattern classification or for pattern discrimination: selection of features merely because they possess the highest variability does not mean that they will necessarily perform optimally in pattern classifiers. Other relevant factors include the compatibility of the scales of the various axes in feature space, and it is only in cases where these originate from obviously compatible parameters that it is likely to be rigorously valid to use the approach without careful normalization: such cases include particularly the multi-spectral and textural analysis applications noted already.

7.7 Concluding remarks

A complete analysis of the methods of SPR must include methods of supervised and unsupervised learning and incorporate the biologically based ANN methodology as part of the full scenario. All methods have their advantages and limitations, and also specific niches and domains of application. The adage 'horses for courses' is appropriate, and indeed it is surprising how much the data varies between one application and another.

It is also interesting how data at one location varies over time. Thus acquisition of a data set and careful analysis of it is of crucial importance to the design of recognition systems. Better by far in a new application area to perform adequate analysis and to let unsupervised learning take its course; and to let validation sets show when training is inadequate. Above all, better to check out as far as possible how well training procedures perform vis-à-vis in-line programmed recognition systems based on conventional structural recognition algorithms. For replacing the human experience blindly by schemes that learn everything directly from the data is a step the boldest system designer must still take warily.

The basic theory of SPR is covered by standard works such as Duda and Hart (1973), Devijver and Kittler (1982) and Jain and Dubes (1988). ANNs have come into wide use relatively recently, following the introduction of the backpropagation algorithm by Rumelhart *et al.* in 1986, and now they are settling into the role of accepted tools for image processing applications. Useful general references on ANNs include Hertz *et al.* (1991), Haykin (1994) and the review article by Hush and Horne (1993). Greenhill and Davies (1994) carries a cautionary tale about the benefits and problems of using ANNs for eliminating noise from images (see also Davies 1997e).

Part 2

Application to food production

This part of the book shows how the generic image processing methodology described in Part 1 relates to a variety of applications in the food industry. The applications covered include baked products, inspection of the cereal grains used in their production, X-ray inspection techniques, agriculture, meat processing and system design. First, the main inspection procedures to be adopted in these application areas are outlined and placed in context.

Chapter 8
Inspection and inspection procedures

8.1 Introduction

Chapter 1 touched on the need for quality control in the production of foodproducts. In this chapter we consider this problem in more detail. Quality control is achieved by inspection procedures which aim to compare the products being produced against previously established standards introduced with this purpose in mind. While the very highest quality is in principle desirable, the technology of the day will not necessarily have reached the stage of permitting products to be assessed to the required level of accuracy, or at least of permitting them to be assessed to this level economically. Thus there is a potential gap between the requirements and what can be provided. However, vision methodology is advancing steadily, as is also the hardware for implementing it in real time: in many cases the stumbling block is on the image acquisition side, and there is still a significant need for more sensitive sensing devices capable of providing high quality X-ray, infra-red, ultrasound or other images which can be used to reliably assess the surface and internal structures of products. Unfortunately, there is also the point that inspection may not be economic even in cases where vision technology has advanced sufficiently: the capability for taking some action on a line is not the same as the managerial decision of implementing it. Let us reconsider the uses to which inspection can be applied which were set out in Chapter 1:

- feedback of information about items on product lines, to control plant
- feedforward of information to reject defective items from product lines
- logging of information to record numbers and characteristics of items on product lines
- general process monitoring to confirm that activity in a manufacturing cell is proceeding normally.

The first of these is typified by control of the thickness of chocolate on chocolate biscuits. Chocolate is an expensive raw material and increasing the thickness to ensure that no 'show-through' appears in a chocolate coating can be costly; thus after careful consideration, and bearing in mind that the basic cost of the vision based feedback loop may be several thousand pounds *plus* the cost of ongoing adjustment and maintenance (e.g. to clean the conveyor and the camera lens, and to replace the lights regularly), this feedback loop may be omitted from the production line. Clearly, one factory manager might regard such an omission as a false economy, especially if he has to supply demanding supermarket chains, though another might regard it as sound economics, especially if he is aiming at rock bottom prices and large sales. We can see these sorts of considerations applying also to the other three items on the above list. For example, rejection of product from a line represents a clear loss which would only be justified if very high quality had to be insisted upon. Similarly, logging of data and general process monitoring appear to be luxuries which many managers might not wish to bother with unless some clear external influence were brought to bear.

Perhaps the most important outside influence is the law, or the fear of litigation. In general, food manufacturers are unlikely to be sued for minor variations in the chocolate or other content of a product, unless of course a minimum content level is printed on a packet. However, manufacturers are expected to maintain 'due diligence' to ensure cleanliness, safety and so on, and these aspects are at their most serious when foreign bodies of various sorts appear in foodstuffs. Every now and again reports appear in the press about a mouse tail appearing in a cereal packet, or glass in baby food, or pieces of metal in packets of frozen food. Some of these instances are no doubt genuine, and it would be very difficult for a manufacturer to check every item rigorously for every possible contaminant. For example, slivers of glass can be quite hard to detect, while small pieces of wood or even mouse droppings can be virtually indistinguishable in X-ray images from the substrate vegetable matter (such as chopped peppers) which is the normal content of a foodpack. Nevertheless, manufacturers are expected to show due diligence, and in the modern factory, if a technological aid is available which can help to ensure safety, it is increasingly incumbent on the manufacturer to make use of it. Here, the march of technology and the march of legislation and public expectation go hand in hand. Inspection processes that are only dreamt about today will be common practice in a year or two's time. With this changing scene, manufacturers will have to be alert to the cost of failing to introduce adequate inspection procedures as well as the cost of doing so. There is also the point that installing inspection equipment and the associated feedback or rejection mechanisms may be quite an expensive business which is most economically carried out when refurbishing a line or building a new one.

Finally, it ought to be remarked that monitoring manufacturing lines can be a necessity in itself: if it is not carried out adequately, outsize products can jam packing machines, and indeed, it is not impossible for undersize products which tend to result in irregularly shaped packets to do the same.

8.2 Phases in the inspection process

The process of inspection involves several major phases. First is the image acquisition phase. In many ways this is the most important phase, since data that has been missed will not be recoverable on a fast moving line, and data that has become distorted or inaccurately represented (e.g. by sensors with too low a resolution) may not be interpreted properly. There is also the problem that a poor lighting system may make the images difficult to analyse, and this may render real-time processing impossible. While it is generally useful to make vision algorithms robust enough to overcome problems of poor image quality, care in the setting up of acquisition systems can make inspection far more reliable and cost-effective and therefore improve viability.

The next phase in the process of inspection is that of object location. In many situations, objects can be quite small and numerous, as for chocolates or beans on a conveyor, and the major part of the computation occurs in locating them accurately: once located, they are relatively straightforward to scrutinize, since only a restricted number of pixels appears within a well-defined region of interest. In fact, the reason why location can be computationally tedious is that an unconstrained search for products has to be undertaken over the entire input space. In addition, even when a product has tentatively been located, it still has to be orientated; at this stage product variability may also be a problem. However, there are some situations where a separate stage of product location may not be necessary. One is where a conveyor is completely covered by numerous small products (as for cereal, pulse or coffee granule inspection) and it can be assumed that products are everywhere: in many such cases it will not be necessary to scrutinize the products individually, but rather to regard them as constituting a continuous texture which is to be examined for homogeneity and other relevant parameters. It is also possible that if many small products appear randomly on a conveyor, though the background shows through in various places, each pixel can be classified as product (acceptable or otherwise) or as background by the same classifier that scrutinizes the product, thereby eliminating the need to have a separate phase for locating it.

The final phase in the process of inspection is that of scrutiny and measurement. Here it is normally necessary for all parts of a product to be examined carefully for distortions or defects of various types. While the product area is often far less than the whole area available within the input image, the fact that each pixel may have to be examined a number of times for different types of defect will sometimes result in this phase requiring quite large amounts of computation. However, some types of hardware processor with a *region of interest* facility can tackle the inspection problem efficiently once the shape to be analysed has reliably been demarcated.

The inspection problem itself involves several important aspects. First, there are size measurements—notably of length and width—though thickness may also have to be checked, and there may also be overt 3D gauging problems (for the most part we simplify the discussion here by ignoring 3D structure). Second, there are shape analysis problems, and the whole object may have to be measured quite carefully to determine whether there are any shape distortions or flaws such as broken corners. Third, there are possible surface defects: we have already referred to lack of chocolate cover on chocolate biscuits, and on fish the batter may be inadequate or of poor texture. Other defects may have to be sought, such as missing blobs of cream or (for a box of chocolates) missing items. These considerations reflect the adequacy of the basic product or any special additives (especially cream, decorative cherries, icing patterns, olives, and so on).

A more serious type of fault is the appearance of foreign objects which should be nowhere near the line, let alone on a product. Here there are a great many possibilities: pieces of metal, rubber, plastic, stone, glass, wood, and so on. Sometimes there is no obvious means whereby the item in question could have arrived on the scene, though in some cases malicious damage cannot be ruled out. However, where people work near the line, it is always possible for objects such as paperclips or rubber bands to drop out of their pockets—a situation that greater rigour can guard against. Shreds of wood can sometimes be knocked off the wooden palettes used to convey packages around factories. But it is also possible for wood to appear (in the form of twigs, etc.[1]) with the raw vegetable source materials that are being processed. Other materials that can appear in this way are pieces of metal and stone, while glass may sometimes be chipped off bottles when these are used for packaging. While rigour in the factory arrangement can overcome these problems most of the time, there is no substitute for inspection at the final packaging stage, as this is the last place on the line where quality can be safeguarded.

Unfortunately, food products are often multiply packaged, first in cellophane, then in packets, then in packs, then in boxes, for example, and

---

[1] Twigs, leaves and other such material constitute a serious problem, and are commonly known in the trade as 'EVM' (extraneous vegetable matter).

a number of monitoring points will be needed if rigorous control of quality is to be achieved. Further back down the line, the entire product may be coated, e.g. with chocolate or batter, so additional inspection points may be needed at such stages too. Clearly, it may be excessively expensive to install and maintain several monitoring points. Thus there is a tendency to maintain a single monitoring point: it could be placed early on to monitor the flow of raw material; later in the line to allow some feedback of temperature or some other relevant parameter; or before the first packaging machine, to monitor carefully what goes into each packet. But however careful the monitoring, there is the possibility that the packaging machines themselves could be responsible for the insertion of tiny pieces of metal, and in any case malicious insertion could occur at any stage until the product leaves the factory. Hence it is the final packaged product that should be subject to the fullest inspection, especially if only one stage of scrutiny can be afforded. Under these circumstances the only possibility would appear to be X-ray inspection. Nowadays, the radiation from X-ray inspection machines can be kept at a very low level, well within the stipulated safety limits for food, but X-ray inspection stations normally cost large sums—not least because of the substantial amounts of shielding and electronic calibration and the interlock mechanisms required to ensure human safety. Minimum figures of some £40,000 are relevant for the most basic commercial X-ray inspection stations incorporating conveyors for food applications, while costs two or three times this figure are not uncommon. These factors mean that great care must be exercised when considering the layout and cost of inspection systems in the food industry.

8.3 Details of the inspection process

In the last section some indication was given of the types of inspection task that may need to be undertaken. Defects could appear under various headings:

- geometric defects
- structural defects
- surface defects
- incidence of foreign objects.

Geometric and structural defects can be examined together, as they involve the right features being at the right positions on products. They may be checked by adherence to idealized templates which are notionally placed over the products as they are located. However, it is sometimes better to search for the specific features in their own right, and to measure their relative positions carefully: a lot depends on the nature of the product. For example, if a biscuit has a rectangular shape and no additives, a template might provide a good check on its shape and at the same time act as a suitable region for textural analysis of its surface appearance. On the other hand, the positions of cherries or strips of icing on a cake might be analysed better by instituting a specific search and measurement procedure as there may be such variability in their locations that a full object template would be subject to significant matching problems.

Surface defects can be detected by thresholding or adaptive thresholding procedures once the product boundaries have been located. However, textured products will have to be tested by appropriate texture analysis procedures within the product boundary regions, as indicated earlier (note that texture analysis procedures tend to *enhance* regions of specific texture, so thresholding will still have to be applied at the final stage). In either case, special lighting may be required, either to illuminate surfaces uniformly or to provide glancing illumination and thereby show up cracks or surface bumps more easily. Here it should be remembered that surface defects may arise from surface relief or surface discoloration, and that the lighting will have to be set up to enhance the type of fault that is to be located. It is also possible that full assessment of surface relief via photometric stereo (Chapter 6) or other means may sometimes be required.

Foreign objects may well make themselves evident while surface defects are being sought, but their potential occurrence may be so serious (as in the case of poisonous moulds or insects in shipments of cereal) that special algorithms have to be devised to detect them using completely unconstrained search. In such cases, specific algorithms designed to detect each individual type of contaminant may have to be employed (see Chapter 10).

Food products have two special characteristics that need to be taken into account during inspection:

1. They are subject to a high degree of variability, both of the raw and of the processed product: no two beans or peas appear exactly the same (in spite of the adage "as like as two peas in a pod"!) and no two current buns look the same (e.g. all the air holes and currents in the buns will have different sizes and positions). This variation in appearance of food products is expected by consumers who actually want individually prepared food items.
2. They are expected to be physically attractive, and indeed the consumer is strongly influenced by the physical appearance when selecting food products and judging their quality.

These factors have enormous importance and mitigate against purely deterministic assessment of food quality. In particular, judgements of food quality have to be partly statistical—as might be expected when, as very frequently happens, surface textures are important to the appearance of a product (this occurs especially with cakes, biscuits and puddings). The other consequence of appearance being an important aspect of perceived food quality is that it is subjective. Nevertheless, there is often considerable agreement between consumers about what constitutes an attractive product. The main problem is thus one of communication and specification: how can an automatic inspection system know when a product is unattractive and reject it? In simple cases, this problem could be tackled by training a neural network or other classifier to make judgements that emulate acknowledged human experts. However, as so much depends on the individual case, we shall not pursue this problem further here: a full solution would no doubt involve assessment of symmetry or even the slight asymmetries which the human eye sometimes finds even more appealing!

Finally, it should be pointed out that the variability that food products are subject to applies with even greater force to the possible defects and foreign objects. Indeed, when searching for foreign objects it has to be borne in mind that they may arise from a variety of sources, may be made out of a variety of materials, may have a variety of sizes and shapes, and may merge so well with their food substrate that even a human operator cannot locate them reliably. This happens particularly when objects such as twigs or elastic bands appear amid a set of randomly chopped peppers in an X-ray image. Ultimately, it is the unpredictability of foreign objects that makes them difficult to find. What is required is for algorithms to be tuned to see the unusual, though it is not at all obvious what this requirement amounts to until it is found that some 'obvious' foreign object has been missed. We shall return to this topic in Chapter 11.

8.4 Lighting schemes

Image analysis has been applied in many areas over the past twenty years, and is particularly important for inspecting products during manufacture. As in many other applications of image analysis, much depends on reliable acquisition of the images that are to be processed. Indeed, it is frequently pointed out that no part of the whole process has greater importance than image acquisition (Batchelor *et al.* 1985; Davies 1997e). There are many reasons for this, but amongst them is the fact that pictorial information that has become degraded cannot in general be restored to its original state, and crucial accuracy once lost cannot be regained; in addition, the ease with which analysis of images can proceed depends to a large extent on how the images have been constructed and in particular on whether a suitable choice of lighting scheme has been made. For example, we may note that a suitable lighting scheme can make the picture data considerably less complex, so that a simple technique such as thresholding can reliably be used to start the process of analysing it. Indeed, this has the further consequence that the processing will take less computational effort, so that the resulting hardware for achieving real-time inspection should be significantly cheaper. Hence care and attention to image acquisition and in particular to the lighting scheme can eventually pay for itself several times over.

Of particular interest in this section is the need to search for blemishes on the surface of materials. Hence, if a region of highly uniform lighting can be set up, then variation in the reflected light can immediately indicate a defect or blemish in the material being inspected. Common materials for which this type of inspection is appropriate include paper, cloth, wood, steel strip, and dough. However, uniform lighting will also be of considerable value in any application where thresholding can be used for binarizing an image, as in optical character recognition, 2D shape analysis, and so on.

The next subsection outlines the principles by which a region of uniform illumination may be produced. Three schemes by which this may be achieved are examined, and design curves which should help with the task of setting up practical lighting schemes for this purpose are presented.

8.4.1 Principles for producing regions of uniform illumination

While initially it may appear to be necessary to illuminate a worktable or conveyor uniformly, a more considered view is that a uniform flat material should *appear* uniform, so that the spatial distribution of the light emanating from its surface is uniform. The relevant quantity to be controlled is therefore the radiance of the surface (light intensity in the image). Following the work of Section 6.3 relating to Lambertian (matt) surfaces, the overall reflectance R of the surface is given by:

$$R = R_0 \mathbf{s}.\mathbf{n} \tag{8.1}$$

where R_0 is the absolute reflectance of the surface and $\mathbf{n},\mathbf{s}$ are respectively unit vectors along the local normal to the surface and the direction of the light source.

Clearly, the assumption of a Lambertian surface can be questioned, since most materials will give a small degree of specular reflection, but in this section we are mainly interested in those non-shiny substances for which equation (8.1) is a good approximation. In any case, special provision normally has to be made for examining surfaces with a significant specular reflectance component. However, notice that the continuous strip lighting systems considered below have the desirable property of very largely suppressing any specular components.

Next we recognize that illumination will normally be provided by a set of lights at a certain height h above a worktable or conveyor. We start by taking the case of a single point source at height h. Supposing that this

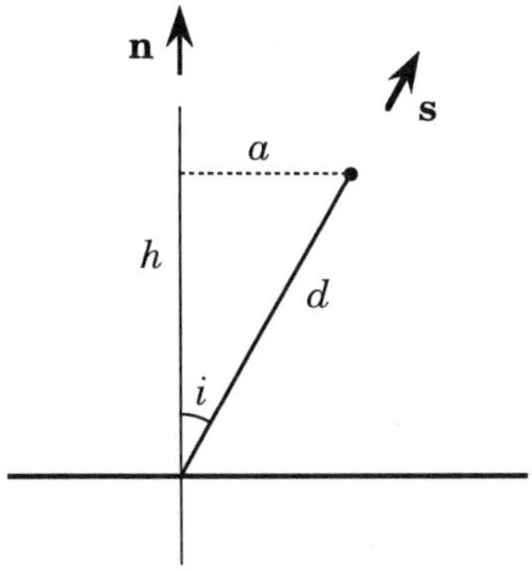

Figure 8.1 Geometry for a single point source illuminating a surface. Here a point light source at a height h above a surface illuminates a general point with angle of incidence i. **n** and **s** are respectively unit vectors along the local normal to the surface and the direction of the light source.

is displaced laterally through a distance a, so that the actual distance from the source to the point of interest on the worktable is d, I will have the general form:

$$I = \frac{c\cos i}{d^2} = \frac{ch}{d^3} \tag{8.2}$$

where c is a constant factor (see Figure 8.1).

Equation (8.2) represents a distinctly non-uniform region of intensity over the surface. However, this problem may be tackled by providing a suitable distribution of lights. A neat solution is provided by a symmetrical arrangement of two strip lights which will clearly help to make the reflected intensity much more uniform (Figure 8.2). We illustrate this idea by reference to the well-known arrangement of a pair of 'Helmholtz' coils—widely used for providing a uniform magnetic field, with the separation of the coils made equal to their radius so as to eliminate the second order variation in field intensity. (Note that in a symmetrical arrangement of coils, all odd orders vanish, *and* a single dimensional parameter can be used to cancel the second order term.)

In a similar way, the separation of the strip lights can be adjusted so that the second order term vanishes (Figure 8.2(b)). There is an immediate analogy also with the second order Butterworth low-pass filter which gives a maximally flat response, the second order term in the frequency response curve being made zero, and the lowest order term then being the fourth order term (Kuo 1966). In fact, the latter example demonstrates how the method might be improved further—by aiming for a Chebychev type of response in which there is some ripple in the pass band, yet the *overall* pass-band response is flatter (Kuo 1966). In a lighting application, we should aim to start with the strip lights not just far enough apart so that the second order term vanishes, but slightly further apart, so that the intensity is *almost* uniform over a rather larger region (Figure 8.2(c)). This reflects the fact that in practice the prime aim will be to achieve a given degree of uniformity over the maximum possible size of region.

In principle it is easy to achieve a given degree of uniformity over a larger region by starting with a given response and increasing the linear dimensions of the *whole* lighting system proportionately. Though valid in principle, this approach will frequently be difficult to apply in practice: for example, it will be limited by convenience and by availability of the strip lights; it must also be noted that as the size of the lighting system increases, so must the power of the lights. Hence in the end we will have only one adjustable geometric parameter by which to optimize the response.

Finally, note that in most practical situations, it will be less useful to have a long narrow working region than one whose aspect ratio is close to

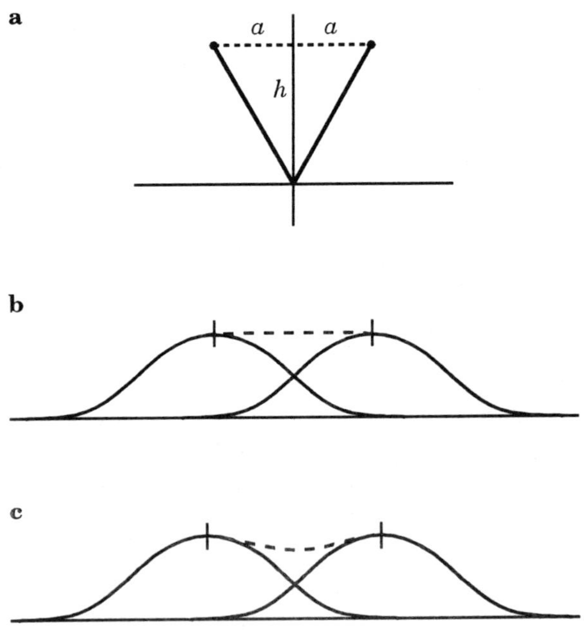

Figure 8.2 Effect of using two strip lights for illuminating a surface. (a) shows two strip lights at a height h above a surface, and (b) shows the resulting intensity patterns for each of the lights; the dotted line shows the combined intensity pattern. (c) shows the corresponding patterns when the separation of the lights is increased slightly.

unity. We shall consider two such cases—a circular ring light and a square ring light. The first of these is conveniently provided in diameters of up to at least 30 cm by commercially available fluorescent tubes, while the second can readily be constructed—if necessary on a very much larger scale—by assembling a set of four linear fluorescent tubes. In this case we take the tubes to be finite in length, and in contact at their ends, the whole being made into an assembly that can be raised or lowered to optimize the system. Thus, these two cases have fixed linear dimensions characterized in each case by the parameter a, and it is h that is adjusted rather than a. To make comparisons easier, we assume in all cases that a is the constant and h is the optimization parameter (Figure 8.3).

8.4.2 *Case of two infinite parallel strip lights*

First we take the case of two infinite parallel strip lights. In this case the intensity I is given by the sum of the intensities I_1, I_2 for the two tubes:

$$I_1(x) = h\int_{-\infty}^{\infty} [(a-x)^2 + (v-y)^2 + h^2]^{-3/2} dv \tag{8.3}$$

$$I_2(x) = I_1(-x) \tag{8.4}$$

Suitable substitutions permit equation (8.3) to be integrated, and the final result is:

$$I = \frac{2h}{(a-x)^2 + h^2} + \frac{2h}{(a+x)^2 + h^2} \tag{8.5}$$

Differentiating I twice and setting $d^2I/dx^2 = 0$ at $x = 0$ eventually (Davies 1997c) yields the maximally flat condition:

$$h = \sqrt{3}a \tag{8.6}$$

However, as noted above, it should be better to aim for minimum *overall* ripple over a region $0 \le x \le x_1$. The situation is shown in Figure 8.4. We take the ripple ΔI as the difference in height between the maximum intensity I_m and the minimum intensity I_0, at $x = 0$, and on this basis the maximum permissible deviation in x is the value of x where the curve again crosses the minimum value I_0.

A simple calculation shows that the intensity is again equal to I_0 for $x = x_1$, where

$$x_1 = (3a^2 - h^2)^{1/2} \tag{8.7}$$

the graph of h versus x_1 being the circle $x_1^2 + h^2 = 3a^2$ (Figure 8.5, top curve). It is interesting that the maximally flat condition is a special case of the new one, applying where $x_1 = 0$.

Further mathematical analysis of this case is difficult: numerical computation leads to the graphs of Figure 8.5. The top curve in Figure 8.5 has already been referred to, and shows the optimum height for selected ranges of values of x up to x_1. Taken on its own this curve would be valueless as the accompanying non-uniformity in intensity would not be

a

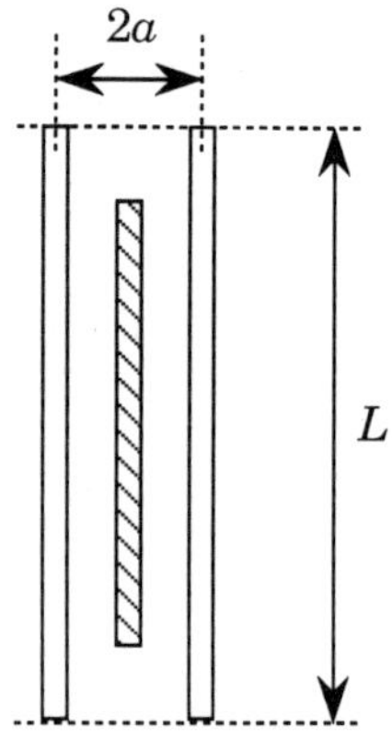

b

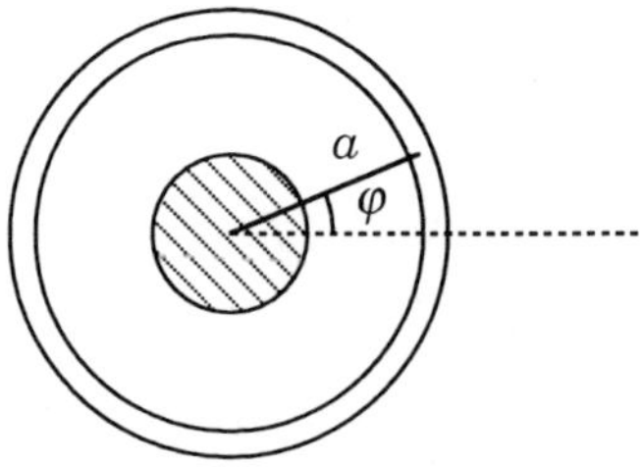

c

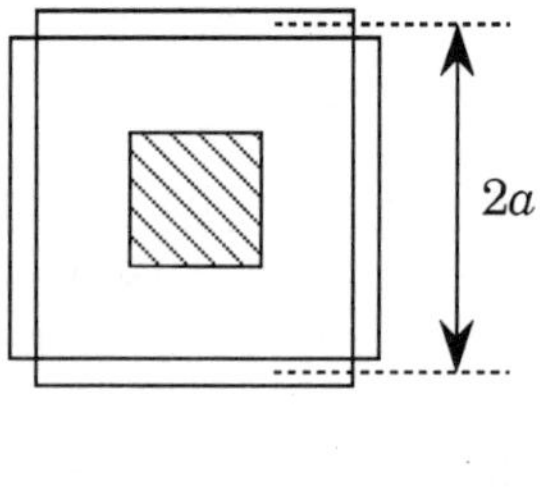

Figure 8.3 Lighting arrangements for obtaining uniform intensity. This diagram shows three arrangements of tubular lights for providing uniform intensity over fairly large regions, shown cross-hatched in each case. (a) shows two long parallel strip lights, (b) shows a circular ring light, and (c) shows four strip lights arranged to form a square 'ring'. In each case height h above the worktable must also be specified.

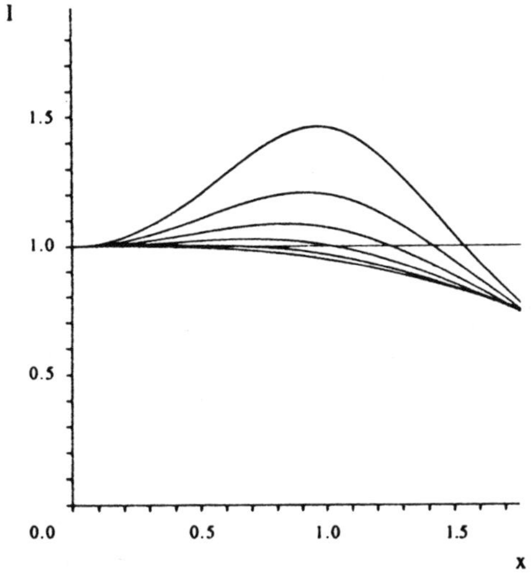

Figure 8.4 Intensity variation for two infinite parallel strip lights. This diagram shows the intensity variation I as a function of the distance x from the centre of symmetry for six different values of h: h increases in steps of 0.2 from 0.8 for the top curve to 1.8 for the bottom curve. The value of h corresponding to the maximally flat condition is $h = 1.732$. x and h are expressed in units of a, while I is normalized to give a value of unity at $x = 0$.

known. This information is provided by the left curve in Figure 8.5. However, for design purposes it is most important first to establish what range of intensities accompanies a given range of values of x, since this information (Figure 8.5, bottom curve) will permit the necessary compromise between these variables to be made. Having decided on particular values of x_1 and ΔI, the value of the optimization parameter h can then be determined from one of the other two graphs: both are provided for convenience of reference (it will be seen that once two of the graphs are provided, the third gives no completely new information). Maximum acceptable variations in ΔI are assumed to be in the region of 20%, though the plotted variations are taken up to ~50% to give a more complete picture; on the other hand, in most of the practical applications envisaged here, ΔI would be expected not to exceed 2–3% if accurate measurements of products are to be made.

The ΔI *v.* x_1 variation varies faster than the fourth power of x_1, there being a very sharp rise in ΔI for higher values of x_1. This means that once ΔI has been specified for the particular application, there is little to be gained by trying to squeeze extra functionality through going to higher values of x_1, i.e. in practice ΔI is the controlling parameter.

8.4.3 *Case of circular and square ring lights*

In the case of a circular ring light, the mathematics is more tedious (Davies 1997c) and it is not profitable to examine it here. The final results are very similar to those for parallel strip lights, and are presented by the graphs in Figure 8.6(a). They would be used for design in the identical manner to that outlined earlier for the previous case.

In the case of a square ring light, the mathematics is again tedious (Davies 1997c), but the results follow the same pattern (Figure 8.6(b)) and warrant no special comment.

8.4.4 *Summary*

Previous work on optical inspection systems has largely ignored the design of optimal lighting schemes. This section has tackled the problem in a particular case of interest—how to construct an optical system that makes a uniform matt surface appear uniformly bright, so that blemishes and defects can readily be detected with minimal additional computation. Three cases have been taken which cover a good proportion of practical lighting schemes, and the design principles described here should be applicable to

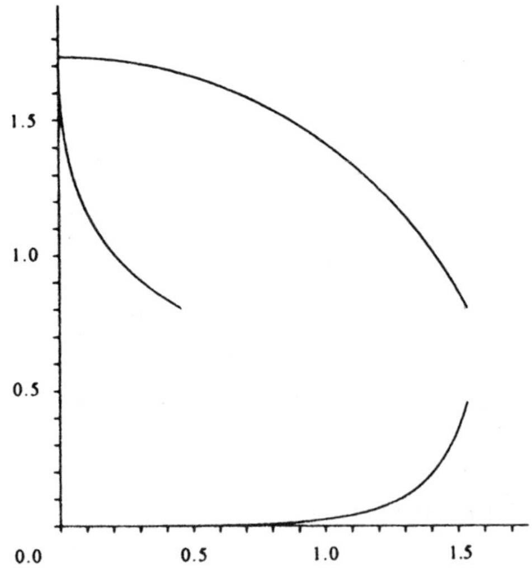

Figure 8.5 Design graphs for two parallel strip lights. Top, h *v.* x_1. Left, h *v.* ΔI. Bottom, ΔI *v.* x_1. The information in these graphs has been extracted from Figure 8.4. In design work, a suitable compromise working position would be selected on the bottom curve, and then h would be determined from one of the other two curves. In practice, ΔI is the controlling parameter, so the left and bottom curves are the important ones, the top curve containing no completely new information.

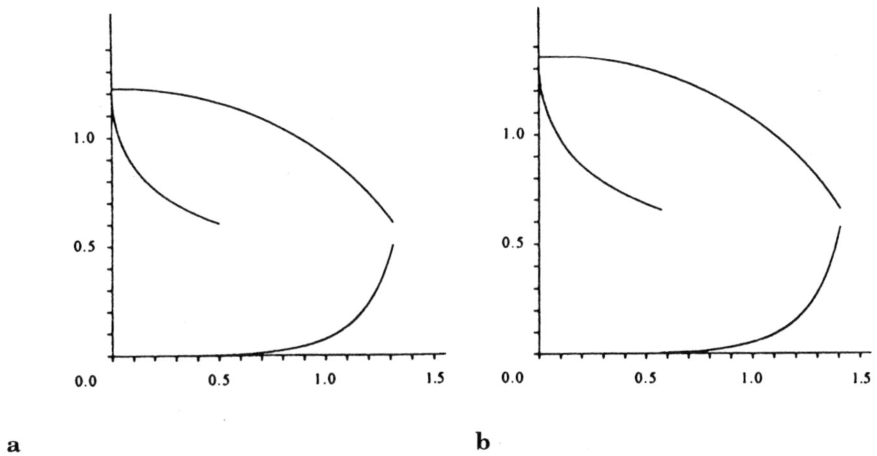

Figure 8.6 Design graphs for a circular ring light (a) and a square ring light (b). Top, h *v.* x_1. Left, h *v.* ΔI. Bottom, ΔI *v.* x_1. For further information relating to these graphs, see the caption to Figure 8.5.

most other schemes that could be employed.

The results in the three cases had to be presented in the form of graphs. In each case, one graph shows the tradeoff between variation in intensity and range of position on the working surface, from which a suitable working compromise can be selected. The other two graphs provide data for determining the optimization parameter (the height of the lights above the working surface).

8.5 Concluding remarks

This chapter has provided an introduction to the inspection work of Part 2 relating to applications of image processing in the food industry. The following chapters cover baked products, inspection of cereal grains, X-ray inspection techniques, agriculture, meat processing and system design. The problem of hardware required for implementation of real-time systems will come up repeatedly, in various guises and forms. These particular strands will be pulled together in Chapter 14, and the whole subject area will be reviewed in Chapter 15. For a wider view of inspection, covering very many aspects of manufacture—not relating especially to food production, the reader is referred to the recent thoroughgoing survey by Newman and Jain (1995).

Chapter 9

Inspection of baked products

9.1 Introduction

In this chapter we shall consider the problems of inspecting baked food products. In particular, we shall take a few key inspection case studies and examine the make-up of viable cost-effective solutions in these cases; this will permit us to comment on available vision techniques and to draw some general lessons; it will also permit us to understand the problems of inspection, determining which have been solved, which are underlying, which are ongoing, and which are the real problems that remain to be tackled. We start in the next section by taking Jaffacake inspection as a basic case study.

9.2 A basic case study: Jaffacake inspection

The particular case study described in this section relates to the inspection of Jaffacakes. This case study is important as it characterizes a whole area of inspection, throwing up a number of crucial problems, and leading to some important lessons. Issues involved include accuracy, robustness, speed of processing and hardware cost, to name but a few.

9.2.1 The problem

Jaffacakes are round shallow cakes which are topped by a small pool of jam and covered with chocolate (see Figure 9.1). When the chocolate has solidified, the products are packed, first into paper packets and then into boxes. Although from the consumer's point of view the total weight of product in a packet may be more important than the exact number of cakes included in it, the size of the individual products is nevertheless crucial: this is because more than about 10% variation in diameter can result in jamming of the packing machine. Hence product diameter is a vital inspection parameter. Likewise, product circularity is important. Next, presence, quantity and placing of the jam must be considered, and it must be checked that the top of the product is completely covered with chocolate. It so happens that chocolate is a highly expensive commodity, and most consumers want a fair ration of it. Thus it is vital to provide just enough chocolate to ensure complete cover, and the inspection system must be sensitive to any deviation from this criterion. Finally, there is a requirement that the whole product should look attractive and inviting. Hence a whole–product quality appearance check has to be included in the inspection system.

9.2.2 The solution

In the particular application considered, the whole inspection process had

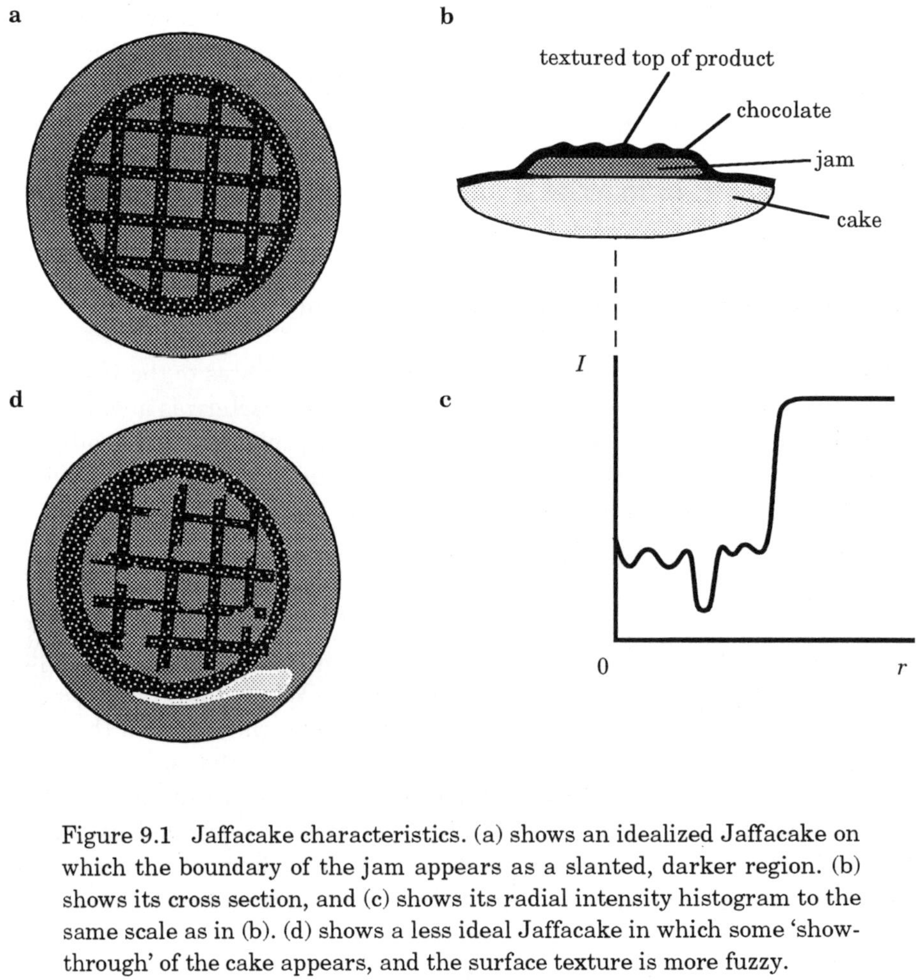

Figure 9.1 Jaffacake characteristics. (a) shows an idealized Jaffacake on which the boundary of the jam appears as a slanted, darker region. (b) shows its cross section, and (c) shows its radial intensity histogram to the same scale as in (b). (d) shows a less ideal Jaffacake in which some 'show-through' of the cake appears, and the surface texture is more fuzzy.

to take place at around 20 products per second, and 100% inspection was required so that quality could be guaranteed. All these factors meant that the design of a hardware system for implementing the inspection algorithms was quite exacting. It also meant that the algorithms had to be made as simple and effective as possible. In this respect the only latitude was in the whole–product quality appearance check, where it would have been preferable to make a specific measure of the textural pattern on the chocolate. In fact, this proved problematic as the texture was coarse and irregular, having only three or four ups and downs in intensity over the flat top of the product (see Figure 9.1). Accordingly, suitable measures of texture included checking that the intensity averages determined by the radial intensity histograms (see below) were within valid limits, and that the textural intensity variation (measured as the distribution of very dark pixels within the chocolate region, found by thresholding) was also within valid limits.

For ease of hardware implementation (and hence cost-effective implementation), extensive use was made of summation processes, including:

1. summation of pixels representing the product area
2. summation of pixels representing chocolate cover
3. summation of pixels representing dark areas within the chocolate region
4. summation of contributions to the radial intensity histogram
5. summation of contributions to the radial intensity histogram correlation function.

Of these, the first two items are self-explanatory. The third has been mentioned already in the context of texture analysis within the chocolate region: however, it actually sums pixels for which the chocolate is slanted away from the horizontal (since it reflects the overhead light sideways), and hence includes the edge of the circular jam region. The fourth item sums and averages the intensities of pixels within concentric rings around the centre of the product (Davies 1984a, 1985): an idealized form of the resultant histogram is shown in Figure 9.1. Although the radial intensity histogram can be assessed directly, it is also assessed indirectly (fifth item) via a correlation function with the radial intensity histogram of an ideal product. This correlation function is used to give an overall quality function for the general appearance of the product.

While this summarizes the product scrutiny and measurement aspects of the inspection process, it is clear from the above description that these depend on identification and accurate location of the centre of the product. In addition, image acquisition has so far been ignored. Below we shall briefly consider image acquisition before going on to consider product location.

9.2.3 Image acquisition

In this application, the product line had a continuously moving conveyor, and it was natural to use a line-scan camera to progressively build up the images. The camera was directly over the line which was also illuminated from above by a set of four symmetrically placed spotlights. All algorithms were made sufficiently robust so that exact placement of the lights was not essential for satisfactory operation, though excessive displacement or failure of a light would result in lower accuracy in the analysis of the grey-scale intensities on the product. Thus catastrophic failure of the inspection process would not occur, and performance would degrade gracefully with loss of lighting function. Because of the problems involved in analysing images containing several products across the width of the conveyor, some of which might be only partially included in any image, it seemed better to go for a very high speed single stream of product, even though this is more exacting with regard to speed of processing. It was decided early on to have only one inspection station at the end of the line where quality control is crucial, rather than having additional stations to check the cake or cake-plus-jam stages, as this would add excessive cost to the overall system: this decision involved judging whether the saving in cost of an additional inspection station would exceed the possible wastage from adding jam or chocolate to an already defective product. Perhaps the main deciding factor was the effectiveness of a final inspection stage in being able to check on all aspects of the product—including presence of jam underneath the chocolate—in a single operation.

9.2.4 Product location

We now move on to the product location operation. Here, three main procedures were considered. In the first, thresholding would lead to binarized images from which individual products could be located by rather trivial operations. However, the frequent occurrence of chocolate on the conveyor meant that these operations became anything but trivial; in addition, thresholding commonly leads to variations in apparent product size unless reasonably sophisticated grey-scale calibration procedures are adopted. This meant that an edge detection-based process was preferable. However, boundary tracking is complicated by the need to link edge segments that peter out in regions of low contrast (especially where plain cake merges with the background conveyor, and where chocolate on the conveyor merges with the chocolate on the product). The boundary tracking approach also gives rise to problems where two Jaffacakes are in contact, or where a Jaffacake overlaps a smear of chocolate on the conveyor. In particular, the centroid of the tracked boundary can be well away from the centre of the product, and thus the centroidal profile (see Chapter 3) is unlikely to be interpretable by any straightforward algorithm. These considerations led to use of the Hough transform (HT) approach for circle centre location (Davies 1984a).

The idea behind the HT is to build up evidence for a particular solution—in this case for the presence of a circle with centre at a given location—while ignoring any irrelevant information pointing to other solutions. As indicated in Chapter 4, this can be achieved by taking each edge point in the input image and the associated information on edge orientation, deducing the position where the centre of a circle would be if the edge were part of the circle boundary, and then accumulating this evidence in the form of a vote at that location: although this procedure requires a knowledge of the radii of the circles, this is not generally a problem with products on a product line. By the time all the edge points have been processed in this way, a set of clusters of votes will have been accumulated in parameter space, together with further votes spread randomly over the parameter space. These extra votes can be ignored, but all the main clusters will constitute peaks which are then located as accurately as possible in order to determine the circle centre locations.

It will be clear from the above description that the HT locates the centres of circles that are present in the image, and that it is not confused by occlusions, breakages, distortions, shadows, noise or other artefacts. It should be noted that this property is intrinsic to the algorithm: in general, it is difficult to add this sort of robustness to a poor algorithm, and this demonstrates the particular advantage of the HT approach.

The HT implementation described above requires an estimate to be made of the orientation of each edge point. This can conveniently be achieved with an accuracy approaching 1° by use of the Sobel operator. It is interesting that this operator functions within a small (3×3) window, and is able to strike an excellent balance between edge localization and edge orientation, while at the same time being computationally highly efficient (Davies 1997e).

9.2.5 Hardware for real-time operation

So far only the algorithmic side of the inspection task has been considered in any detail, though it was hinted above that use of summations in several of the scrutiny and measurement operations would make the algorithms straightforward to implement in dedicated hardware for real-time operation. In fact, a similar situation applies to the Sobel operator, which involves two 3×3 convolution operations, again requiring simple summations.

Davies and Johnstone (1986, 1989) performed a detailed analysis of the whole inspection task, and showed the tradeoffs that exist between hardware and software. Their general conclusion was that there is a strict priority for implementing it in hardware, based on the *c/t* (hardware cost divided by software execution time) parameter for each individual software module, modules with low values of *c/t* having to be implemented before those of higher *c/t*. For the Jaffacake inspection task, the highest priority for hardware implementation was the relatively slow 3 × 3 convolution operations for edge detection, whereas the scrutiny and measurement functions, which are single (1 × 1) pixel operations, had significantly lower priority. Curiously, the HT operations, which are often regarded as highly computation intensive, could easily be performed on the host computer, without the help of dedicated hardware: here a saving in time was achieved by clearing the parameter space globally only once, and reclearing it by *un*writing the votes that had been written during the previous object location operation. Logging of results and instituting product rejection were far from requiring hardware implementation (Davies and Johnstone give a number of criteria for determining when to stop implementing software modules in dedicated hardware). These considerations led to an implementation which could inspect Jaffacakes at up to 12 products per second—more than half the line rate—in a single machine costing around £12,000. More advanced technology, e.g. bit-slice devices or DSPs (cf. Edmonds and Davies 1991; Davies *et al.* 1995), would by now significantly reduce this figure, but the methodology and the optimization procedure outlined by Davies and Johnstone apply to all such implementations.

9.2.6 Discussion

One of the important things that is learnt from this case study is that product location can be far more computation intensive than product scrutiny and measurement. This is because product location is a search process which requires object recognition irrespective of the position and orientation of the object. On the other hand, scrutiny can in principle be regarded as placing a template over the product and performing a pixel-by-pixel matching operation. In fact, product location intrinsically takes $O(N^2n^2Q)$ operations[1], where the image is an $N \times N$ array of pixels and the product template is an $n \times n$ array, and Q orientations of the product can be distinguished (see Chapter 4). On the other hand, product scrutiny takes just $O(mn^2)$ operations, m is the number of times each pixel must be examined to check for surface details, such as jam, chocolate, texture, etc. (in the Jaffacake case m was five, as indicated earlier. Thus design of inspection algorithms tends to involve significantly more attention being paid to the determination of product location (and particularly the speed

---

[1] Note that use of the HT has the effect of limiting n to three in this expression if a 3 × 3 pixel edge detection operator is used.

with which this can be achieved) than to the precise measurements to be made on the product.

Next, it is relevant to ask how much would change if products other than Jaffacakes were being inspected. The answer to this question clearly varies with a great many factors including:

- the shape of the product
- the appearance of the product
- the particular properties to be measured.

Let us first consider appearance. At this point we assume that the products to be inspected are still round food products. Then the main difference when compared with the Jaffacake case study lies in (a) the radius and (b) the intensity profile. In fact, the Jaffacake algorithm could cope with wide variations in both of these, without additional programming, by learning from examples of suitable close-to-ideal products. Thus it was fairly trivially adaptable from Jaffacakes to round biscuits, pizzas, and so on.

In principle, the Jaffacake algorithm was also able to cope with accurately machined round parts, such as pistons, but in practice accuracy had to be augmented, as the original algorithm was optimized for high speed with only the moderate accuracy demanded by highly variable food products (for which variations in diameter of up to 3% are normally acceptable). Again, variations in the shapes of the product necessitated significant redesign of the algorithm, though the same general design principles—such as use of HTs—could clearly be adopted.

Space does not permit a detailed discussion of the possible variations in the design of the algorithm, though the following particular techniques are worthy of note:

- use of HTs for straight line detection (Chapter 4)
- use of HTs for ellipse detection (Chapter 4)
- use of sampling to speed up circle centre location (Section 9.4.5)
- use of a hybrid sequential/parallel HT approach for accurate (sub-pixel) circle centre location (Davies 1988a)
- use of HTs, median differencing, and lateral histograms for corner and small hole detection (Paler *et al.* 1984; Davies 1988d, 1997e)
- use of a 'truncated median'-based iterative procedure for efficient highly accurate (sub-pixel) location of HT peaks (Davies 1992b)
- use of robust statistics for highly accurate (sub-pixel) location of HT peaks (Hampel *et al.* 1986; Meer *et al.* 1991).

Extensive tests of these techniques were made, and they all have particular characteristics which recommend them for different inspection and other applications. However, incorporating them all into the same software package does not seem to be worthwhile, since the techniques are still evolving, and new ones are being developed: above all, their suitability depends heavily on the type of image data to be analysed. As yet the machine vision community is still apparently at a technique gathering stage rather than at a definitive algorithm prescription stage. (This is of course contrary to what most published papers announcing new techniques and packages generally claim! However, the truth is that most researchers work with highly specialized datasets which make it difficult to generalize to different applications.)

Overall, the adaptation of the Jaffacake algorithm to different products and to different applications must take account of the following fundamental parameters:

- the size and shape of the product
- the intensity, texture and colour of the product
- the structural complexity of the product
- the variability of the product
- the required accuracy of location and measurement
- the noise level in the image data
- the degree of 'clutter' in the background.

These data-dependent factors must be balanced by the following algorithm design factors:

- sensitivity (in a signal-to-noise ratio sense)
- accuracy
- robustness
- adaptability
- speed
- cost of real-time hardware implementation.

All these factors have to be optimized for cost-effective practical application, but it must be recognized that improving one factor may result in others being degraded. For example, improving accuracy may demand an increase in resolution, which will inevitably increase execution time. Similarly, improving speed by algorithmic means may necessitate use of sampling procedures, which will in turn cut down sensitivity and accuracy (accuracy is normally dependent upon the signal-to-noise ratio that can be achieved) (Davies 1988b). Finally, cost of real-time hardware will inevitably be dependent upon the speed that is required. It will be clear from this discussion that all six factors are closely linked, and that any optimization will in fact take place in a 6D space. In such a space, there will be places where there is no good solution, and in practice this will mean that cost-effective applications cannot be produced until some algorithmic breakthrough occurs: any algorithm will have its own region of applicability within the optimization space, but will only come into its own, being better than all other available algorithms, in an even more restricted region of this space. See Section 14.3 for further discussion of this point.

By way of example, we cite the lateral histogram approach to small hole location (Davies 1997e). This is excellent for non-cluttered images containing relatively few holes and little noise: indeed, it is amongst the fastest and most effective practical ways of finding small holes in such images, being considerably faster than the HT. However, the time taken for checking the potential hole locations increases as $O(p^2)$ for p holes, so the method eventually ceases being competitive for this purpose, and it is then necessary to revert to use of the HT or other methods. Operation of the method also depends strongly upon the size of the holes. Clearly, the lesson here is that each algorithm has its own area of application, and this depends both on the task and on the signal data and on the remaining data in the image. Thus algorithm specification is a complex task, while algorithm design is generally even more complex, since it may not be clear until an algorithm has been designed and tested whether a given specification is realisable. See Chapter 14 for further elaboration of these points.

9.3 Case study: inspection of cream biscuits

Although it might be thought that foodproduct inspection usually takes on the form outlined in the previous section, we here describe a separate approach which proved useful in a case of cream biscuit inspection (Figure 9.2). Each cream biscuit consisted of two wafers separated by a layer of cream. In such cases, not only can the usual defects such as breakage occur, but also excessive cream can ooze out from between the wafers; in addition, the wafers can be misaligned. It is necessary to inspect for both occurrences, since they result in changes in overall size and shape and can be the cause of packaging problems. They also detract from product appearance, and hence tend to inhibit sales.

When faults like those indicated above occur, they disrupt the normal boundary of the product and make it difficult to locate and orientate it accurately. It is therefore better to locate the product accurately and reliably via the decorative 'docker' holes on the surface of the top wafer, and then to place a template over it to determine whether the boundary falls within acceptable tolerance limits. Thus an algorithm which locates products from their holes is required. In fact, the maximal clique approach (Chapter 4) has long been a paradigm for this type of problem.

As outlined in Chapter 4, the basic idea of the maximal clique approach is to label all the features (i.e. holes) in the image and the idealized template, and to consider what happens if hypothetical matches are made between the two. In the resulting match graph, the largest groups of mutually compatible feature assignments—the maximal cliques—are sought and taken to represent the most likely identifications of objects.

As noted in Chapter 4, the maximal clique computation is exponential in the number of features, and the HT provides a cost-effective approach to the task of object location. The basis of this technique is to take a reference

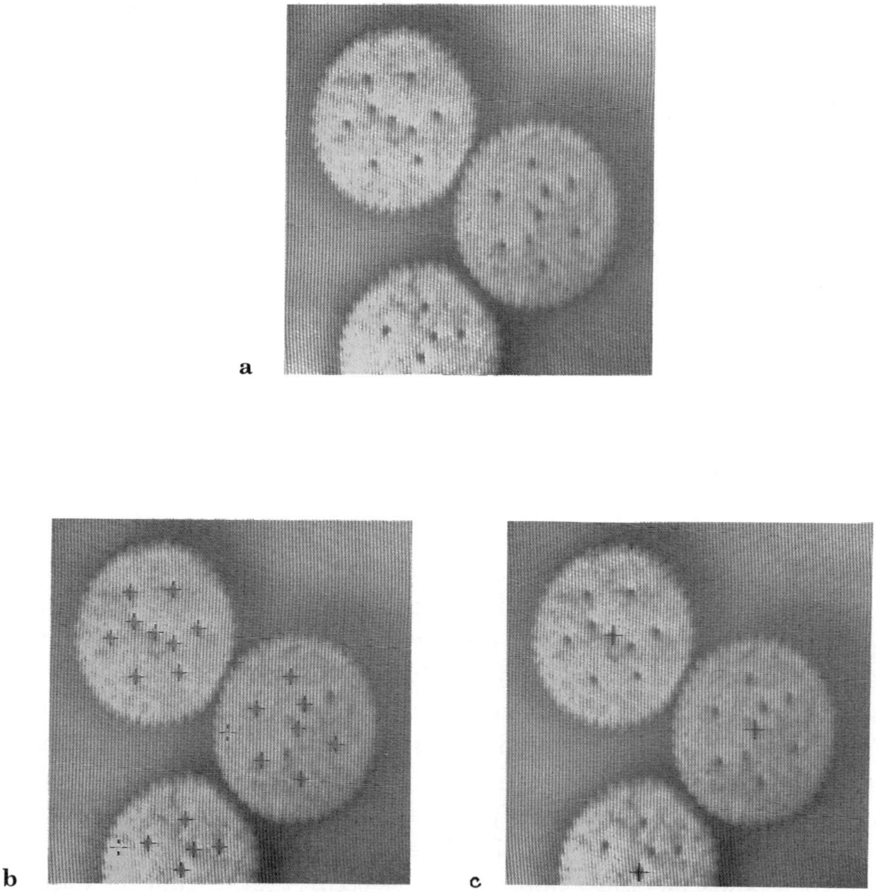

Figure 9.2 Location of cream sandwich biscuits from their 'docker' holes. (a) Original image. (b) Result of applying a simple fairly fast hole detection algorithm which misses two of the holes and finds some non-existent holes. (c) Reliable biscuit location using the Hough transform. Note also that partial occlusion of the third biscuit does not prevent its location.

point P in the object, and to use the HT to place votes at P according to the separation distances of every pair of features on the object. In the present case, it turns out that cream biscuits usually possess quite a number of symmetrically placed docker holes, and advantage can be taken of this to minimize the computational load: this is achieved by taking P to be at the centre of symmetry of the pattern of docker holes (Figure 9.2). While for very simple patterns the maximal clique approach could possibly give a quicker solution than the HT, for cases where there are six or more features the HT approach is far more efficient (Davies 1991a, 1992a): this follows from the rapid deterioration in performance of the maximal clique's computation characteristic.

In contrast with the above approach, Edmonds and Davies (1991) adopted a more conventional boundary detection method in the case of rectangular cream biscuits with a chocolate coating where no hole features were visible.

9.4 Short case studies of baked product inspection

This section includes a few short case studies which will broaden out the scope of the chapter: in particular they will show other factors that are important in the production of baked products. Figures 9.3 and 9.4 give some idea of the problems involved in the first three of these case studies.

Figure 9.3 Slice of bread. Notice the single large bubble amongst the usual distribution of small bubbles.

Figure 9.4 Loaf of bread showing a pronounced 'overspill'.

9.4.1 Description of the textural appearance of bread crumb (Bertrand et al. 1992)

This paper is concerned with the scientific assessment of bread crumb in slices of bread loaves. Haar masks of various sizes and shapes, but typified by the form:

$$\begin{bmatrix} 1 & -1 \\ -1 & 1 \end{bmatrix}$$

were used to analyse the texture pattern. Of the resulting 66 Haar coefficient values, each averaged over the whole image, seven qualitative groups were created and discriminant analysis (see Section 7.2) carried out to identify the bread treatment from its crumb texture. About 80% of the samples were correctly identified, but loaf-to-loaf variation was found to be important. The methodology was felt to be useful for quality control of the bread-making process.

9.4.2 Development of an objective crumb-grain measurement (Rogers et al. 1995)

This study was aimed at the production of a rugged and versatile system for objective bread crumb-grain measurement. The method used was to apply Fourier analysis to 64 × 64 pixel subimages of bread slices and obtain 2D power spectral density distributions; each was then thresholded and fitted at its outer reaches by an ellipse whose eccentricity and orientation gave an estimate of the fineness and elongation of the crumb texture pattern. (For ideal isotropic crumb patterns, the ellipses would be circles.) The fineness parameters showed correlations with those of human experts of around 80%, though there was less consistency over the elongation scores (humans appear to judge elongation by a variety of intuitive methods).

9.4.3 Automated inspection of bread and loaves (Batchelor 1993)

This paper considers the inspection of loaves of bread from three major points of view—shape analysis, analysis of the crumb texture, and detection of foreign bodies within the loaf. The paper makes many suggestions about how image processing might be used to help with these tasks—for example: (1) analysis of the 3D shape of the loaf with the aid of structured lighting, analysis of the 2D shapes of slices using convex hulls, and Radon transforms (similar to Hough transforms) to estimate the positions of straight sides; (2) image texture analysis via filtering, thresholding and morphological processing followed by counting of black areas to provide information on the crumb texture; (3) use of directional lighting to accentuate the crumb texture; and (4) use of X-rays with grey-scale morphology to locate foreign objects. It is emphasized that shape analysis is complicated by the fact that loaves have ill-defined shapes and low dimensional tolerance, so that loaf shapes over a wide range are acceptable. Nevertheless, the following shape factors are expected to be important in determining the acceptability levels—maximum dimensions, shape symmetry, tapering, crevices, overspill (Figure 9.4) and concavities.

9.4.4 Fast boundary extraction of biscuits (Wu and Rodd 1991)

This paper describes an approach to fast boundary extraction of objects that may be touching. The approach is rule-based and is able to separate touching objects from information gained during boundary tracking. The paper reports application in the food industry for segmentation of biscuits on a moving conveyor.

9.4.5 Fast location of biscuits by line sampling (Davies 1987)

This paper describes work done by the author to find the centres of circular biscuits significantly more rapidly than for a conventional Hough transform, while retaining as far as possible the robustness of that approach. The best solution appeared to be to scan along a limited number of horizontal lines in the image recording and averaging the x-coordinates of mid-points of chords of any objects, and repeating the process in the vertical direction to complete the process of centre location. The method was successful and led to speedup factors as high as 25 in practical situations, some of the speedup being attributable to use of a 1D edge detector. Figure 9.5 shows how progressively lower sampling levels still permitted the biscuits to be located: accuracy and reliability were found to be subject to graceful degradation—a necessary criterion for success with such a method. Figure 9.6 shows the limitations of the approach in the need for fairly careful selection of the gradient threshold, to avoid problems due to the intensity variations of the chocolate coating.

9.4.6 Location of cherries on cup cakes (Chan and Batchelor 1993)

This article outlines how cup cakes can be inspected for proper placement of a cherry within the icing region on the top of each cake. The outermost pastry region, the central icing region and the cherry are segmented and identified with the aid of colour: then if the cherry is absent or if it overlaps the pastry region (the latter situation is detectable by an AND operation) the cake is judged to be unacceptable.

9.5 Concluding remarks

The case studies presented in this chapter show that application of machine vision to food processing in general, and inspection in particular, is more complex than might at first be imagined. A main conclusion is that algorithms must be designed from the outset to cope with the nasty realities that can arise in real images—particularly shadows, glints, occlusion, noise and clutter, to name but a few. They must also be designed to cope with the realities of actual products—specifically, variability in size, shape, intensity, texture or colour, whether these are designed variations or chance occurrences which could cause an algorithm to become confused or to fail dramatically.

In addition, in real food applications the algorithms will have to operate in real time[2]. Normally, this means that special fast dedicated hardware is required to implement them, as indicated in Section 9.2.5. While certain board-based solutions to the hardware implementation problem (such as convolution processors) are intrinsically attractive, system optimization may necessitate less hardware than a crude calculation would indicate, because there are certain operations that can profitably be carried out on the host processor. We shall see in Chapter 11 that this concept can

[2] Nevertheless, in a recent study (Whitworth and Alava 1999), bubble size distributions in bread doughs were obtained and analysed in a thoroughgoing way to obtain accurate results and increased knowledge of the processes involved in breadmaking, real-time operation not being a major requirement.

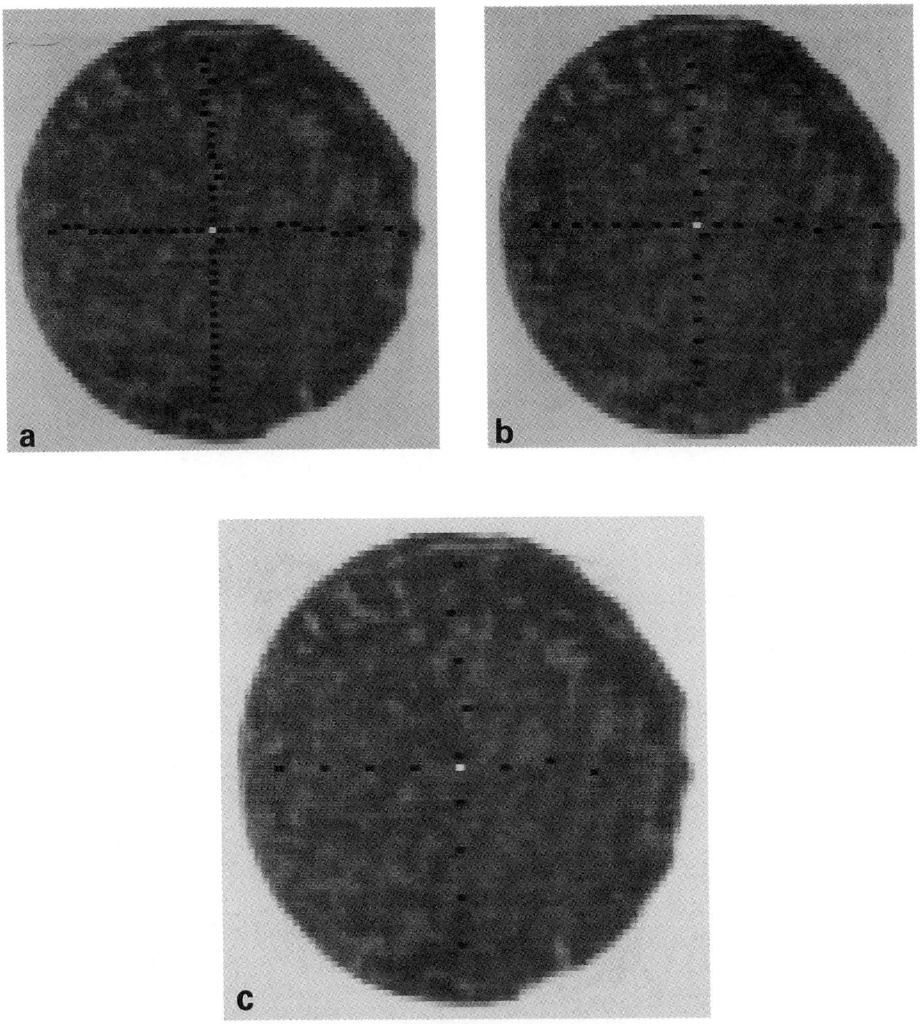

Figure 9.5 Examples of successful object location using the chord bisection algorithm for the same initial image, using successive step sizes of 2, 4 and 8 pixels. The black dots show the positions of the horizontal and vertical chord bisectors, and the white dots show the computed centre locations.

be extended, by opting for a flexible software-based solution: thus the dedicated hardware that is needed may usefully be limited.

Finally, in this chapter we are already starting to see the ubiquity of shape analysis as a tool for object recognition and characterization. This is so whether we are looking at a flat surface with visible light, a 3D object with curved surfaces by visible or structured light, or the inside of a 3D object as a 2D X-ray image. We should first note the variety of modalities for forming the image, and second note the wide choice of algorithms that

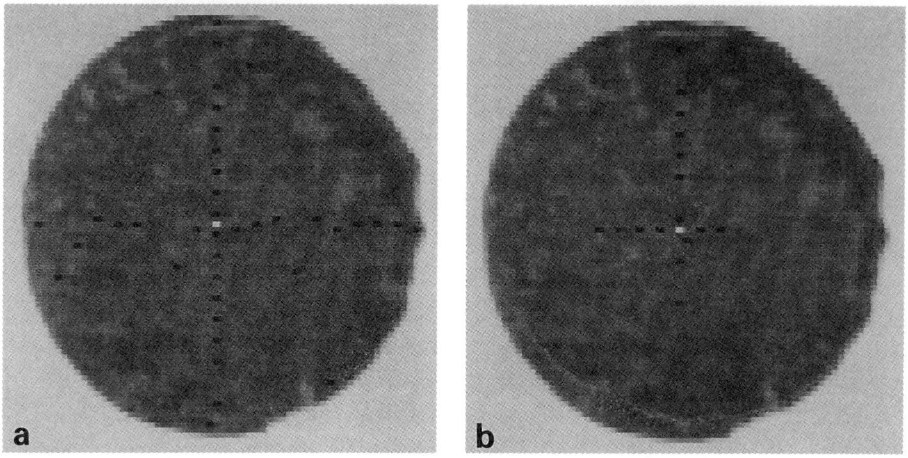

Figure 9.6 Effect of misadjustment of the gradient threshold. (a) Effect of setting the threshold too low, so that surface texture confuses the algorithm. (b) Loss of sensitivity on setting the threshold too high.

are available for shape analysis: the list of such algorithms referred to already includes:

- boundary pattern analysis
- convex hulls
- ellipse modelling
- Hough transforms
- morphological processing.

In addition, surface examination and characterization by texture analysis has also been found important. In the following chapters, we shall find the list of methods for shape and surface characterization growing, in line with what has already been suggested in Part 1.

It must not be imagined that the case studies in this chapter exhaust the possibilities for food processing and inspection—or for novel ways of analysing the resulting images. Suffice it to mention three more applications: first, the robotic chocolate decoration scheme developed by Cronshaw as early as 1982; second, the use of learning algorithms to judge the quality of zig-zag icing patterns on cakes (Naim *et al.* 1994); third, the measurement of colour baking curves and their use to control baking with the aid of neural networks (Yeh *et al.* 1995; Hamey and Yeh 1996).

Chapter 10
Cereal grain inspection

10.1 Introduction

Cereal grains are amongst the most important of the foods we grow. A large proportion of cereal grains is milled and marketed as flour: this is then used to produce bread, cakes, biscuits and many other commodities. Cereal grains can also be processed in a number of other ways (such as crushing), and thus they form the basis of many breakfast cereals. In addition, there are some cereal grains, or cereal kernels, which are eaten with a minimum of processing: rice is an obvious member of this category—though whole wheat and oat grains are also consumed as decorative additives to 'granary' loaves.

Wheat, rice and other cereal grains are shipped and stored in huge quantities measured in millions of tons, and a very large international trade exists to market these commodities. Transport also has to be arranged between farmers, millers, crushers and the major bakers and marketers. Typically, transit by road or rail is in relatively small loads of up to twenty tons, and grain depots, warehouses and ports are not unlikely to receive lorries containing such consignments at intervals ranging from 20 minutes to as little as 3 minutes. All the necessary transportation and storage results in grains being subject to degradation of various sorts: damage, moulds, sprouting, insect infestation, and so on. In addition, the fact that grains are grown on the land and threshed implies the possibility of contamination by rodent droppings, stones, mud, chaff and foreign grains. Finally, the quality of grain from various sources will not be uniformly high, and varietal purity is an important concern.

These factors mean that ideally, the grain that arrives at any depot should be inspected for a good many possible causes of degradation. This chapter is concerned with grain inspection. In the space of one chapter we shall not aim to cover all possible methods and modes of inspection. Indeed, this would be impossible as the subject is moving ahead quite fast: not only are inspection methods evolving rapidly, but the standards against which inspection is carried out are also evolving quite quickly. To some extent the improvement of automatic inspection methods and the means of implementing them efficiently in hardware are helping to drive the process onwards. We shall first explore the situation with the aid of three main case studies plus some shorter ones, and then we shall look at the overall situation.

The first case study involves the examination of grains to locate rodent droppings and moulds such as ergot. The second case study will consider how grains may be scrutinized for insects such as the saw-toothed grain beetle. The third case study is concerned with inspection of the grains themselves. However, this case study will be more general, and will be less involved with the scrutiny of individual grains than with how efficiently they can be located: this is an important factor when lorry loads of grains are arriving at depots every few minutes—leading to the need to sample of the order of 300 grains per second. As remarked in Chapter 8, object location can involve considerably more computation than object scrutiny and measurement.

10.2 Case study: location of dark contaminants in cereals

As noted above, there is a demand for grain quality to be monitored before processing to produce flour, breakfast cereals and a myriad of other derived products. So far automated visual inspection has been applied mainly to the determination of grain quality (Ridgway and Chambers 1996), with concentration on the determination of varietal purity (Zayas and Steele 1990; Keefe 1992) and the location of damaged grains (Liao *et al.* 1994). However, relatively little attention has been paid to the detection of contaminants. In fact, there is the need for a cheap commercial system which can detect insect infestations and other important contaminants in grain—while not being confused by 'permitted admixture' such as chaff and dust (up to ~2% is normally permitted). The inspection work described in this case study (Davies *et al.* 1998a) pays particular attention to the detection of certain important non-insect contaminants. Relevant contaminants in this category include rodent (especially rat and mouse) droppings, moulds such as ergot, and foreign seeds. (It should be remarked that ergot is poisonous to humans, so locating any instances of it is of especial importance.) In this case study the substrate grain is wheat and foreign seeds such as rape would be problematic if present in too great a concentration.

It will be clear that many of the potential contaminants for wheat grains (and for grains of similar general appearance such as barley or oats) are quite dark in colour. This means that thresholding is the most obvious approach for locating them. However, there are a number of problems, in that shadows between grains, dark patches on the grains, chaff and other admixture components, together with rapeseeds, appear as false alarms, so further recognition procedures have to be invoked to distinguish between the various possibilities. As a result, the thresholding approach is not eventually as attractive as might *a priori* have been thought (Figure 10.1(a, b)). This problem is exacerbated by the extreme speeds of processing required in real applications. For example, a successful device for monitoring lorry-loads of grain might well have to analyse a 3 kg sample of grain in 3 minutes (the typical time between arrival of lorries at a grain terminal), and this would correspond to some 60,000 grains having to be monitored for contaminants in that time. This places a distinct premium on rapid, accurate image analysis.

In this case study we concentrate on monitoring grain for rodent droppings. As indicated above, these types of contaminant are generally darker than the grain background, but cannot simply be detected by thresholding since there are significant shadows between the grains, which themselves often have dark patches. In addition, the contaminants are speckled because of their inhomogeneous constitution and because of lighting and shadow effects. In spite of these problems, the contaminants are identifiable by human operators because they are relatively large and their shape follows a distinct pattern (e.g. an aspect ratio of three or four to one). Thus it is the combination of size, shape, relative darkness and speckle that characterizes the contaminants and differentiates them from the grain substrate.

Designing efficient rapidly operating algorithms to identify these contaminants is something of a challenge. The obvious route is via mathematical morphology: apply an erosion to eliminate the small shadows around the grains, and then dilate to consolidate the contaminants. Unfortunately, erosion can lead to elimination of the contaminants as well as the grain features, because of speckle on the contaminants and also because parts of the contaminants are actually relatively light. We consider the problem in more detail in the next subsection.

10.2.1 Application of morphological and non-linear filters to locate rodent droppings

As indicated above, the obvious approach to the location of rodent droppings is to process thresholded images by erosion and dilation. (Erosion followed by dilation is normally termed 'opening', but we refrain from calling it that here, as we wish to retain the generality of not insisting on exactly the same number of erosion and dilation operations: in inspection it is the final recognition that must be emphasized, rather than some idealized morphological process.) In this way, shadows between grains, and discoloration of grains would be eliminated by the erosions, and the shapes and sizes of the contaminants restored by the subsequent dilations. The

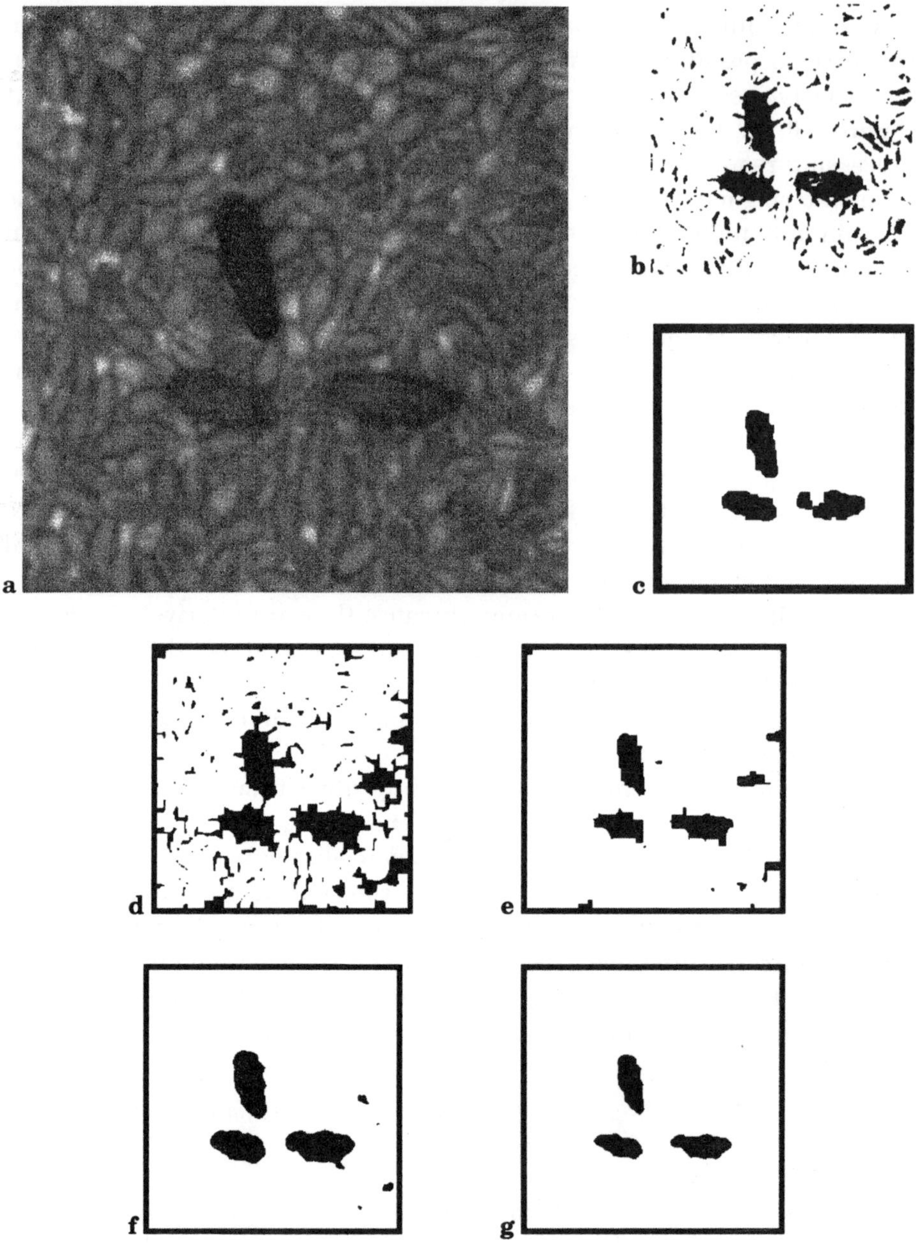

Figure 10.1 Effects of various operations and filters on a grain image. (a) Grain image containing several contaminants (rodent droppings). (b) Thresholded version of (a). (c) Result of erosion and dilation on (b). (d) Result of dilation and erosion on (b). (e) Result of erosion on (d). (f) Result of applying 11×11 median filter to (b). (g) Result of erosion on (f). In all cases, 'erosion' means three applications of the basic 3×3 erosion operator, and similarly for 'dilation'.

effect of this procedure is shown in Figure 10.1(c). Note that the method has been successful in eliminating the shadows between the grains, but has been decidedly weak in coping with light regions on the contaminants. Remembering that while considerable uniformity might be expected between grains, the same cannot be said about rodent droppings, whose size, shape and colour vary quite markedly. Hence the erosion–dilation schema has limited value, though it would probably be successful in most instances. Accordingly other methods of analysis were sought (Davies *et al.* 1998a).

The second approach is to attempt to make good the deficiency of the previous approach by ensuring that the contaminants are consolidated even if they are speckled or light in places. Thus an attempt was made to apply dilation before erosion. The effect of this approach is shown in Figure 10.1(d). Notice that the result is to consolidate the shadows between grains even more than the shapes of the contaminants. Even when an additional few erosions are applied (Figure 10.1(e)) the consolidated shadows do not disappear, and are of comparable sizes to the contaminants. Overall, the approach is not viable, as it creates more problems than it solves (the number of false positives far exceeds the number of false negatives, and similarly for the total areas of false positives and false negatives). One possibility is to use the results of the earlier erosion–dilation schema as 'seeds' to validate a subset of the dilation–erosion schema. However, this would be far more computation intensive and the results would clearly not be especially impressive (see Figure 10.1(c, e)). Instead, a totally different approach was adopted.

The new approach was to apply a large median filter to the thresholded image, as shown in Figure 10.1(f). This gives good segmentation of the contaminants, retaining their intrinsic shape to a reasonable degree, and suppresses the shadows between grains quite well. In fact the shadows immediately around the contaminants enhance the sizes of the latter in the median filtered image, while some shadows further away are consolidated and retained by the median filtering. It was found that a reasonable solution is to perform a final erosion operation (Figure 10.1(g)): this eliminates the extraneous shadows and brings the contaminants back to something like their proper size and shape: though the lengths are slightly curtailed, this is not a severe disadvantage. Overall, the median filtering–erosion schema gave easily the greatest fidelity to the original contaminants, while being particularly successful at eliminating other artefacts (Davies *et al.* 1998a).

Finally, although median filtering is intrinsically more computation intensive than erosion and dilation operations, the first concern is to find the most effective operations that are possible, and the limitations of the most obvious approaches: the work described in this case study is orientated at effectiveness rather than immediate real-time implementation. In any case, many methods have been devised for speeding up median filtering (Davies 1997e), and it is known that this can be done efficiently, and that a variety of hardware platforms are available for performing it rapidly, so it is not an approach which should be eliminated out of hand. We shall disregard the speed aspect in the remainder of the case study.

10.2.2 Appraisal of the various schemas

In the previous section we showed that two basic schemas which aimed in the one case to consolidate the background and then the contaminants and in the other to consolidate the contaminants and then the background failed because each interfered with the alternate region. It appeared that the median-based schema avoided this problem because it was able to tackle both regions at once without prejudicing either. In fact, it is interesting to speculate that the median filter is acting in this case as an analytical device which calculates the final result in a single rigorous stage, thus avoiding any prejudice in the final outcome.

It is useful to examine the situation more closely. The grain substrate can be regarded as forming a background texture, and the speckled contaminants (which also have variable lightness) can be regarded as having a foreground texture which is to be segmented from the background texture. Unfortunately, these textures are quite complex and large-scale, so if for example Laws' masks were used to segment them, the sizes of the masks would need to be very large and the associated smoothing masks would have to be even larger (see Chapters 5 and 11). This consideration makes conventional texture analysis procedures less attractive in this application: they would also be highly computation intensive. The importance of morphological techniques is that they are not computation intensive, but on the contrary they are simple inexpensive procedures. Likewise, a single median filtering operation is affordable, even if it has to be backed up by an erosion operation.

We next examine in more detail why the median filtering–erosion schema is better than the other ones tested in this particular application, and consider whether and under what conditions the situation would be generalisable to other applications.

We can model the basic situation as shown in Figure 10.2(a). Perhaps surprisingly, a 1D model is suitable. First, taking the four basic dimensions k, l, m, n for a strictly periodic set of textures, the model shows that if $k > m$, it is best to start by eroding by $m/2$, as this eliminates the background artefacts without causing irreversible deterioration of the foreground. Second, if $n > l$, it is best to start by dilating by $l/2$ as this consolidates the foreground without causing irreversible deterioration of the background. With either schema the real problem starts when the

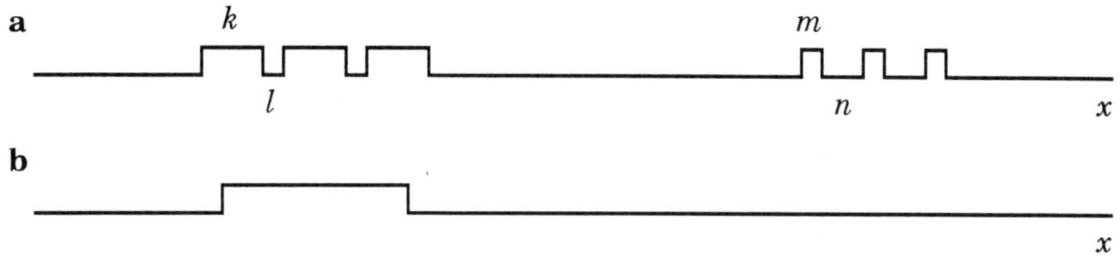

Figure 10.2 Model 1D intensity patterns and the result of applying a median filter. (a) Model intensity pattern with contaminant near left and a set of shadows near right. (b) Effect of applying a 1D median filter with a width of 7 pixels in the case $k = 3, l = m = 1, n = 2$. Note the elimination of the shadow artefacts, and the consolidation of the contaminant.

textures are not strictly periodic, and there is a distribution of values of k, l, m, n. In that case neither schema necessarily wins, and only detailed analysis and tests can determine which gives fewer false classifications. (There is also the problem of whether false positives are less costly than false negatives, and by how much, but we shall not consider this further here.)

The median filtering based schema does not rely on considerations of the precise separations of adjacent dark–light and light–dark transitions, or vice versa, but instead computes averages over several such separations: in fact, it works out and operates on the total dark and total light over a set distance—the width of the window. Thus within the contaminant it regards speckle as light spots which can be made good by sufficient other evidence of contaminant nearby. Similarly, in the background it regards grain shadows as dark holes which can be eliminated by sufficient other evidence of light background nearby. In the model of Figure 10.2(a), the median operation taken over a window $w = 2k + l$ immediately judges in a systematic way that the k and l regions constitute contaminant (though it will clearly ignore the very outside of the contaminant), and ignores the m and n regions (Figure 10.2(b)). It is also clear that this will apply for other textures for which the distributions of values of k, l, m, n are significantly more variable. The crucial point is that the median window w must be wide enough to ignore random dark regions whose integrated width of darkness is D, and this is achieved when $w > 2D$. Indeed, we now see how the system can be optimized—by adjusting w to lie midway between twice the minimum width for a contaminant and twice the maximum integrated width of any possible background dark regions.

Overall, this section has explained why the median-based schema is a more reliable indicator of the presence of contaminant than either of the pure morphologically based schemas which were also examined.

10.2.3 Summary

This case study has considered a problem involved in the inspection of grain for dark contaminants typified by rodent droppings. False alarms due to shadows between and discolorations on the grains have to be eliminated without introducing false negatives. Obvious morphological schemas (erosion–dilation and dilation–erosion) were found to have significant deficiencies which were almost entirely overcome by a median filtering–erosion schema (Davies *et al.* 1998a). Although the latter schema involves more computation, this should not be a practical limitation and in any case the additional rigour it affords justifies the approach. Perhaps more important, the way the schema operates is fully understood. In particular, it is able to cope with variability in the separations of dark and light areas in foreground and background regions, whereas earlier morphological approaches work in such a way that the exact separations between dark–light and light–dark transitions matter and immediately determine success or failure of these techniques. This means that when textural matter is involved (textures being characterisable statistically rather than absolutely), the standard morphological approach is limited whereas the median-based approach can provide a viable solution.

Overall, it seems that the usual types of morphological filter are not powerful enough to identify certain types of object, and that more complex processing strategies are required which combine them with other filtering modalities to give the best results.

10.3 Case study: location of insects

As indicated in the first case study, there is the need for a cheap commercial system which can detect a variety of contaminants in grain. The present case study pays particular attention to the need to detect insects. It is worth noting that insects present an especially serious threat, because they can multiply alarmingly in a short span of time, so greater certainty of detection is vital in this case. This meant that a highly discriminating method was required for locating adult insects (Davies *et al.* 1998b).

Not surprisingly, thresholding initially seemed to be the approach offering the most promise for locating insects, which appear dark relative to the light brown colour of the grain. However, early tests on these lines showed that a good many false alarms would result from chaff and other permitted admixture, from less serious contaminants such as rapeseeds, and even from shadows between, and discolorations on, the grains themselves (Figure 10.3). These considerations led to use of the linear feature detection approach for detecting small adult insects. This proved possible because these insects appear as dark objects with a linear (essentially bar-shaped) structure; hence attempts to detect them by applying bar-shaped masks led ultimately to a linear feature detector which had good sensitivity for a reasonable range of insect sizes. Before proceeding further, we consider the problems of designing linear feature detector masks.

10.3.1 The vectorial strategy for linear feature detection

In Section 4.3 we described how a pair of simple but unusual 3 × 3 masks could in principle be used for line segment detection. In fact the masks L_0,

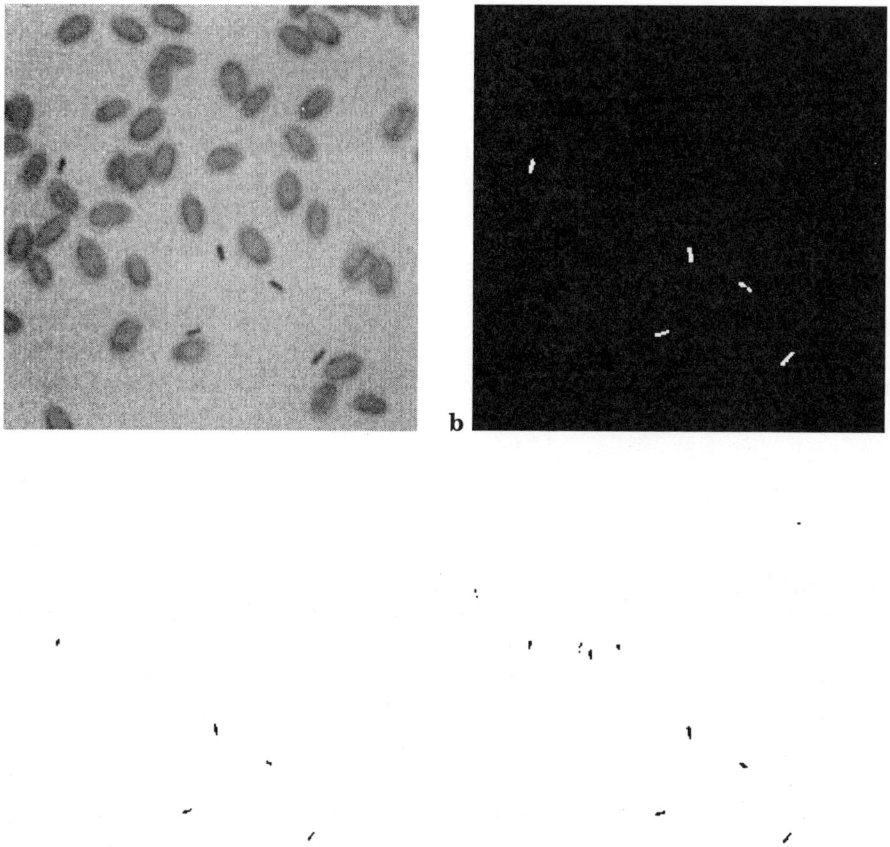

a b

c d

Figure 10.3 Insects located by linear feature detector. (a) Original image. (b) Result of applying linear feature detection algorithm. (c) Result of selecting optimum (minimum error) threshold level on (a). (d) Effect of small increase in threshold. © 1998 EURASIP

L_{45} presented there have to be augmented as follows to give the most accurate results:

$$L_0 = A\begin{bmatrix} 0 & -1 & 0 \\ 1 & 0 & 1 \\ 0 & -1 & 0 \end{bmatrix} \qquad L_{45} = B\begin{bmatrix} -1 & 0 & 1 \\ 0 & 0 & 0 \\ 1 & 0 & -1 \end{bmatrix} \tag{10.1}$$

Next, we have to select appropriate values of the coefficients A and B. Applying the above masks to a window with the intensity pattern

a	b	c
d	e	f
g	h	i

leads to the following responses at 0° and 45°:

$$g_0 = A(d + f - b - h) \tag{10.2}$$

$$g_{45} = B(c + g - a - i) \tag{10.3}$$

and hence to the following formulae for effective line contrast and estimated orientation:

$$g = (g_0^2 + g_{45}^2)^{1/2} \tag{10.4}$$

$$\theta = \frac{1}{2}\arctan(g_{45}/g_0) \tag{10.5}$$

Detailed calculations were carried out by Davies (1997a) for the case of a thin line of width w passing through the centre of the 3 × 3 window, pixel responses being taken in proportion to the area of the line falling within each pixel. However, for lines of substantial width, theory proved intractable and simulations were needed to determine the outcome. What was remarkable about the results was the high orientation accuracy that occurred for $w = 1.4$, in the case when $B/A = 0.86$, which gave a maximum error of just 0.4°. In what follows we will not be concerned to obtain a high orientation accuracy, but will be content to use equation (10.4) to obtain an accurate estimate of line contrast using just two masks, thereby saving on computation. The reason for this is the need to attain high sensitivity in the detection of image features that correspond to insect contaminants, rather than to pass results on to high-level interpretation algorithms, as orientation is not an important output parameter in this application.

10.3.2 Designing linear feature detection masks for larger windows

When larger windows are used, it is possible to design the masks more ideally to match the image features that have to be detected, because of the greater resolution then permitted. However, there are many more degrees of freedom in the design, and there is some uncertainty as to how to proceed. The basic principle (Davies *et al.* 1998b) is to use masks whose profile matches the intensity profile of a linear feature around a ring of radius R centred on the feature, and at the same time follows a particular mathematical model—namely an approximately sinusoidal amplitude variation. For a given linear feature of width w, the sinusoidal model will achieve the best match for a thin ring of radius R_0 for which the two arc lengths within the feature are each one quarter of the circumference $2\pi R_0$ of the ring. Simple geometry (Figure 10.4) shows that this occurs when:

$$2R_0 \sin(\pi/4) = w \tag{10.6}$$

$$\therefore \quad R_0 = w/\sqrt{2} \tag{10.7}$$

The width ΔR of the ring should in principle be infinitesimal, but in practice, considering noise and other variations in the data set, ΔR can validly be up to about 40% of R_0. The other relevant factor is the intensity profile of the ring, and how accurately this has to map to the intensity profile of the linear features to be located in the image. In many

Figure 10.4 Geometry for application of a thin ring mask. Here two quarters of the ring mask lie within, and two outside, a rectangular bar feature.

applications, such linear features will not have sharp edges, but will be slightly fuzzy, and the sides will have significant and varying intensity gradient. Thus the actual intensity profile is quite likely to correspond reasonably closely to a true sinusoidal variation. Masks designed on this basis proved close to optimal when experimental tests were made (Davies *et al.* 1998b). Figure 10.5 shows masks that resulted from this type of design process for one specific value of R (2.5 pixels).

10.3.3 Application to cereal inspection

The main class of insect which was targeted in this study (Davies *et al.* 1998b) was *Oryzaephilus surinamensis* (saw-toothed grain beetle): insects in this class approximate to rectangular bars of about 10×3 pixels, and masks of size 7×7 proved to be appropriate for identifying the pixels on the centrelines of these insects. In addition, the insects could appear in any orientation, so potentially quite a large number of template masks might be required: this would have the disadvantage that their sequential implementation would be highly computation intensive and not conducive to real-time operation. In view of this, it was natural to employ the vectorial approach to template matching, in which two orthogonal masks would be used as outlined above instead of the much larger number needed when using a more conventional approach.

The output of the linear feature detector was taken to give the definitive insect signal and was thresholded to make decisions on the presence of any insects. This approach proved adequate for the location of small adult insects (Davies *et al.* 1998b). Although it resulted in a small false negative rate of around 1 percent, all the cases observed (from a total set of 300 insects) corresponded to specific identifiable problems, such as insects being viewed end on or being partly obscured by grains, or (in one instance) another species of insect, a grain weevil, emerging from a grain!

The procedure also led to a small proportion of false alarms in the proximity of dark grain boundaries, and especially in the region of shadow

·	·	−1	−2	−1	·	·
·	·	−1	−2	−1	·	·
1	1	·	·	·	1	1
2	2	·	•	·	2	2
1	1	·	·	·	1	1
·	·	−1	−2	−1	·	·
·	·	−1	−2	−1	·	·

·	·	−1	·	1	·	·
·	−2	−2	·	2	2	·
−1	−2	·	·	·	2	1
·	·	·	•	·	·	·
1	2	·	·	·	−2	−1
·	2	2	·	−2	−2	·
·	·	1	·	−1	·	·

Figure 10.5 Typical (7×7) linear feature detection masks.

© 1998 EURASIP

between two touching grain boundaries. Thus a specific detector was required to post-process the object field and eliminate such cases. There remained a much smaller number of false alarms due to chaff and other artefacts which simulated the appearance of insects: clearly any method of pattern recognition is unable to distinguish such artefacts from the intended type of object if the visual resemblance is too close. In this case, the only way of improving the recognition rate would have been to increase resolution by a factor of at least two, in which case the speed of recognition would have been reduced by a factor of about four (or equivalently the cost of any fast hardware accelerator would have been increased by a similar factor). Such system optimization issues are beyond the scope of this case study. We merely remark that the linear feature detector procedures described here seem to meet the needs for a highly effective preliminary screening of the input images. Finally, in contrast with the case of edge detection, a suitable line detection strategy is to concentrate on the negative-going line-enhanced image: bright lines do not correspond to any relevant types of contaminant.

10.3.4 Summary and discussion

This case study has shown how the vectorial strategy for designing line segment detection masks can be extended to larger windows, thereby permitting line segments—and, in particular, insects—to be detected and orientated efficiently with just two masks in a far greater number of applications than the original formulation[1]. This work uses a theoretically based strategy which is backed up by experimental tests to ensure that it operates well in practical cases. Specifically, the masks have to be adapted to detect particular linear features (a) by adjusting their radii to match the widths of the linear features which are modelled as rectangular bars of constant width w; and (b) by increasing sensitivity without seriously affecting discriminability by employing ring masks of substantial size. As for the earlier 3×3 masks described by Davies (1997a), a full theoretical analysis is difficult, but the situation is well understood, and experimental tests verify the effectiveness of this type of linear feature detector.

The types of linear feature detector described in this section should be useful in quite different aspects of the food industry: for example, they could be useful for locating thin branches which are to be subjected to automatic pruning.

Next, it should be remarked that the linear feature detector approach will not be the only valid one for searching for insects in grain images: an interesting approach using discriminant analysis has been developed by

---

[1] In more recent work insects have been located with even greater efficiency using isotropic masks (Davies *et al.* 1999); however, space does not permit a full account of this work to be given here.

Zayas and Flinn (1998). However, it is targeted primarily at a different type of insect, *Rhyzopertha dominica* (the lesser grain borer beetle): thus no data is currently available for a direct comparison of the two methods. Finally, an interesting appraisal of different ways of finding pests (insects and mites) in food appears in Chambers (1997).

10.4 Case study: high speed grain location

It has already been mentioned several times in this monograph that object location often requires considerable computation, as it involves unconstrained search over the image data: as a result the whole process of automated inspection can be slowed down significantly, and this can be of crucial importance in the design of real-time inspection systems. Indeed, if the scrutiny of particular types of object requires quite simple dimensional measurements to be made, the object location routine can be the bottleneck. This case study is concerned with the high speed location of objects in 2D images, a topic on which relatively little systematic work has been carried out—at least on the software side—though many studies have been made on the design of hardware accelerators for image processing. As hardware accelerators represent the more expensive option, this case study concentrates on software solutions to the problem, and then specializes it to the case of cereal grains in images.

10.4.1 Extending an earlier sampling approach

In an earlier project, the author had the problem of finding the centres of circular objects such as coins and biscuits significantly more rapidly than for a conventional Hough transform, while retaining as far as possible the robustness of that approach (Davies 1987). The best solution appeared to be to scan along a limited number of horizontal lines in the image recording and averaging the x-coordinates of mid-points of chords of any objects, and repeating the process in the vertical direction to complete the process of centre location. The method was successful and led to speedup factors as high as 25 in practical situations—in particular for the rapid location of round biscuits (Chapter 9).

In the present project (Davies 1998), extreme robustness was not necessary, and it seemed worth finding how much faster the scanning concept could be taken. It was envisaged that significant improvement might be achieved by taking a minimum number of sampling points in the image rather than by scanning along whole lines.

Suppose that we are looking for an object such as that shown in Figure 10.6(a), whose shape is defined relative to a reference point R as the set of pixels $A = \{\mathbf{r}_i : i = 1 \text{ to } n\}$, n being the number of pixels within the object. If the position of R is $\mathbf{x}_\mathrm{R}$, pixel i will appear at $\mathbf{x}_i = \mathbf{x}_\mathrm{R} + \mathbf{r}_i$. This

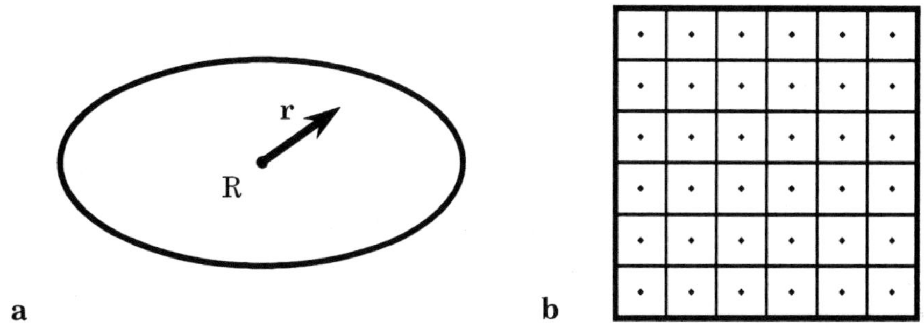

Figure 10.6 Object shape and method of sampling. (a) Object shape, showing reference point R and vector $\mathbf{r}$ pointing to a general location $\mathbf{x}_R + \mathbf{r}$. (b) Image and sampling points, with associated tiling squares.

© 1998 EURASIP

means that when a sampling point $\mathbf{x}_s$ gives a positive indication of an object, the location of its reference point R will be $\mathbf{x}_R = \mathbf{x}_s - \mathbf{r}_i$. Thus the reference point of the object is known to lie at one of the set of points $U_R = \bigcup_i(\mathbf{x}_s - \mathbf{r}_i)$, so knowledge of its location is naturally incomplete. Indeed, the map of possible reference point locations has the same shape as the original object, but rotated through 180°—because of the minus sign in front of $\mathbf{r}_i$. Furthermore, the fact that reference point positions are only determined within n pixels means that many sampling points will be needed, the minimum number required to cover the whole image clearly being N/n, if there are N pixels in the image. This means that the optimum speedup factor will be $N/(N/n) = n$, as the number of pixels visited in the image is N/n rather than N (Davies 1997b).

Unfortunately, it will not be possible to find a set of sampling point locations such that the 'tiling' produced by the resulting maps of possible reference point positions covers the whole image without overlap. Thus there will normally be some overlap (and thus loss of efficiency in locating objects) or some gaps (and thus loss of effectiveness in locating objects). Clearly, the set of tiling squares shown in Figure 10.6(b) will only be fully effective if square objects are to be located.

However, a more serious problem arises because objects may appear in any orientation. This prevents an ideal tiling from being found. It appears that the best that can be achieved is to search the image for a maximal *rotationally invariant* subset of the shape, which must be a circle, as indicated in Figure 10.7(a). Furthermore, as no perfect tiling for circles exists, the tiling that must be chosen is either a set of hexagons or, more practically, a set of squares. This means that the speedup factor for object location will be significantly less than n, though it will still be substantial.

10.4.2 Application to grain inspection

A prime application for this technique is that of fast location of grains on a conveyor in order to scrutinize them for damage, varietal purity, sprouting, moulds, etc. Under these circumstances it is best to examine each grain in isolation: specifically, touching or overlapping grains would be more difficult to cope with. Thus the grains would need to be spread out with at most 25 grains being visible in any 256×256 image. With so much free image space there would be an intensive search problem, with far more pixels having to be considered than would otherwise be the case. Hence a very fast object location algorithm would be of especial value.

Wheat grains are well approximated by ellipses in which the ratio of semi-major (a) to semi-minor (b) axes is almost exactly two. The deviation is normally less than 20%, though there may also be quite large apparent differences between the intensity patterns for different grains. Hence it seemed worth using this model as an algorithm optimization target. First, the (non-ideal) $L \times L$ square tiles would appear to have to fit inside the circular maximal rotationally invariant subset of the ellipse, so that $\surd 2L = 2b$, i.e. $L = \surd 2b$. This value should be compared with the larger value $L = (4/\surd 5)b$ which could be used if the grains were constrained to lie parallel to the image x-axis—see Figure 10.7(b) (here we are ignoring the dimensions $2\surd 2b \times \surd 2b$ for optimal *rectangular* sampling tiles).

Another consequence of the difference in shape of the objects being detected (here ellipses) and the tile shape (square) is that the objects may be detected at several sample locations, thereby wasting computation (see Section 10.4.1). A further consequence of this is that we cannot merely count the samples if we wish to count the objects: instead we must relate the multiple object counts together and find the centres of the objects. This also applies if the main goal is to locate the objects for inspection and

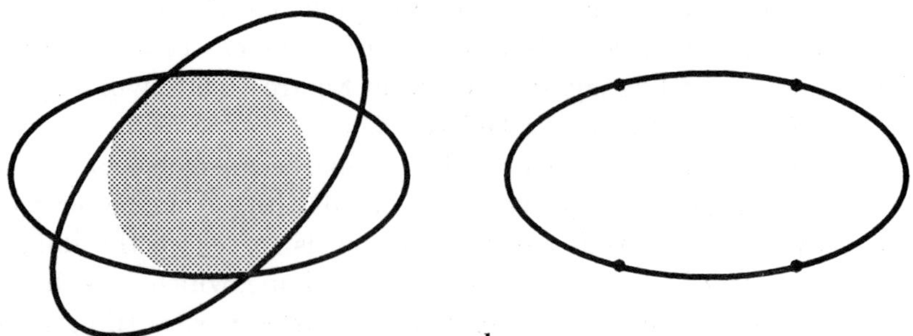

a b

Figure 10.7 Geometry for location of ellipses by sampling. (a) Ellipse in two orientations and maximal rotationally invariant subset (shaded). (b) Horizontal ellipse and geometry showing size relative to largest permitted spacing of sampling points. © 1998 EURASIP scrutiny. In the present case, the objects are convex, so we only have to look along the line joining any pair of sampling points to determine whether there is a break and thus whether they correspond to more than one object. We shall return later to the problem of systematic location of object centres.

For ellipses, it is relevant to know how many sample points could give positive indications for any one object. Now the maximum distance between one sampling point and another on an ellipse is $2a$, and for the given eccentricity this is equal to $4b$ which in turn is equal to $2\sqrt{2}L$. Thus an ellipse of this eccentricity could overlap three sample points along the x-axis direction if it were aligned along this direction; alternatively, it could overlap just two sample points along the 45° direction if it were aligned along this direction, though it could in that case also overlap just one laterally placed sample point. In an intermediate direction (e.g. at an angle of around arctan 0.5 to the image x-axis), the ellipse could overlap up to five points. Similarly, it is easy to see that the minimum number of positive sample points per ellipse is two. The possible arrangements of positive sample points are presented in Figure 10.8(a).

Fortunately, the above approach to sampling is over-rigorous. Specifically, we have insisted upon the sampling tile being contained within the ideal (circular) maximal rotationally invariant subset of the shape. However, what is required is that the sampling tile must be of such a size that all possible orientations of the shape are allowed for. In the present example the limiting case that must be allowed for occurs when the ellipse is orientated parallel to the x-axis, and it must be arranged that it can just pass through four sampling points at the corners of a square, so that on any infinitesimal displacement, at least one sampling point is contained within

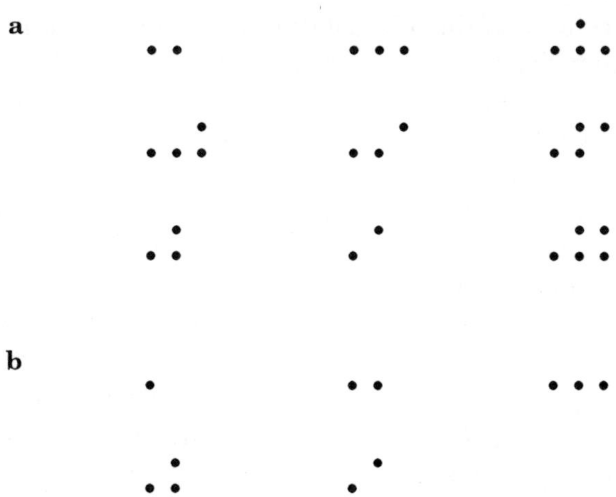

Figure 10.8 Possible arrangements of positive sampling points for ellipse, (a) with $L = \sqrt{2}b$, and (b) with $L = (4/\sqrt{5})b$. © 1998 EURASIP

it. For this to be possible it can be shown that $L = (4/\sqrt{5})b$, the same situation as already depicted in Figure 10.7(b). This leads to the possible arrangements of positive sampling points shown in Figure 10.8(b)—a distinct reduction in the average number of positive sampling points, which leads to useful savings in computation (the average number of positive sampling points per ellipse is reduced from ~3 to ~2).

Object location normally takes considerable computation because it involves an unconstrained search over the whole image space, and in addition there is normally (as in the ellipse location task) the problem that the orientation is unknown. This contrasts with the other crucial aspect of inspection, that of object scrutiny and measurement, in that relatively few pixels have to be examined in detail, requiring relatively little computation. Clearly, the sampling approach outlined above largely eliminates the search aspect of object location, since it quickly eliminates any large tracts of blank background. Nevertheless, there is still the problem of refining the object location phase. One way of approaching this problem is to expand the positive samples into fuller regions of interest and then to perform a restricted search over these regions. For this purpose we could use the same search tools that we might use over the whole image if sampling were not being performed. However, the preliminary sampling technique is so fast that this approach would not take full advantage of its speed. Instead we could use the following procedure.

For each positive sample, draw a horizontal chord to the boundary of the object, and find the local boundary tangents. Then use the chord–tangent technique (join of tangent intersection to mid-point of chord: Davies 1997e) to determine one line on which the centre of an ellipse must lie. Repeat this for the all positive samples, and obtain all possible lines on which ellipse centres must lie. Finally, deduce the possible ellipse centre locations, and check each of them in detail in case some correspond to false alarms arising from objects which are close together rather than from genuine self-consistent ellipses. Note that in cases where there is a single positive sampling point, another positive sampling point has to be found (say $L/2$ away from the first).

In fact, a significantly faster approach called the *triple bisection* algorithm has recently been developed (Davies 1998). Draw horizontal (or vertical) chords through adjacent vertically (or horizontally) separated pairs of positive samples, bisect them, join and extend the bisector lines, and finally find the mid-points of these bisectors (Figure 10.9). (In cases where there is a single positive sampling point, another positive sampling point has to be found, say $L/2$ away from the first.) The triple bisection algorithm has the additional advantage of not requiring estimates of tangent directions to be made at the ends of chords, which can prove inaccurate when objects are somewhat fuzzy, as in many grain images. The result of applying this technique to an image containing mostly well-

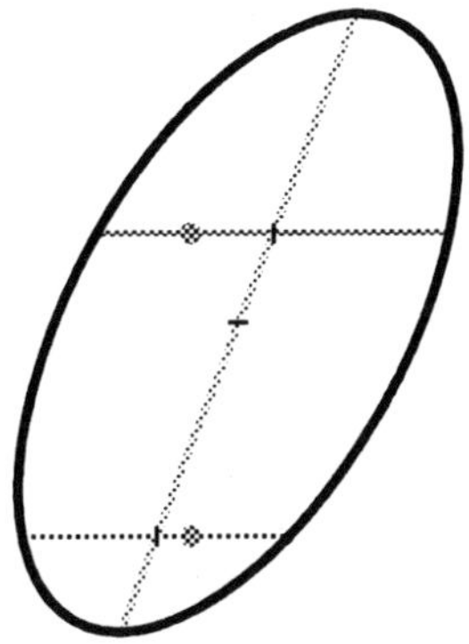

Figure 10.9 Illustration of triple bisection algorithm. The round spots are the sampling points, and the short bars are the midpoints of the three chords, the short horizontal bar being at the centre of the ellipse.
© 1998 EURASIP

separated grains is shown in Figure 10.10: this illustrates that the whole procedure for locating grains by modelling them as ellipses and searching for them by sampling and chord bisection approaches is a viable one. In addition, the procedure is very fast, as the number of pixels that are visited is a small proportion of the total number in each image.

Finally, we show why the triple bisection algorithm presented above is appropriate. First note that it is correct for a circle, for reasons of symmetry. Second, note that in orthographic projection, circles become ellipses, straight lines become straight lines, parallel lines become parallel lines, chords become chords, and midpoints become midpoints. Hence choosing the right orthogonal projection to transform the circle into a correctly orientated ellipse of appropriate eccentricity, the midpoints and centre location shown in the diagram of Figure 10.9 must be validly marked. This proves the algorithm. (For a more rigorous algebraic proof, see Davies 1999.)

10.4.3 Summary

This case study has studied sampling strategies for the rapid location of objects in digital images. Motivated by the success of an earlier line-based sampling strategy (Davies 1987), it has shown that point samples lead to the minimum computational effort when the 180°-rotated object shapes form a perfect tiling of the image space. In practice imperfect tilings have to be used, but these can be extremely efficient, especially when the image intensity patterns permit thresholding, the images are sparsely populated with objects, and the latter are convex in shape. An important feature of the approach is that detection speed is *improved* for larger objects, though naturally exact location involves some additional effort. In the case of ellipses, the latter process is considerably aided by the triple bisection algorithm.

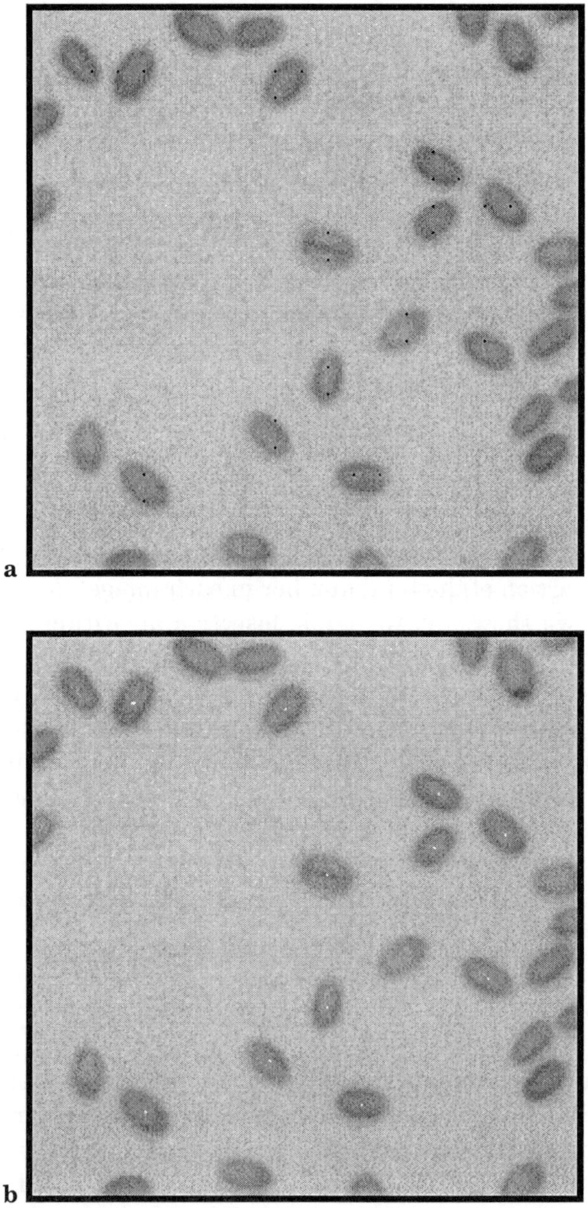

Figure 10.10 Image showing grain location using the sampling approach. (a) Sampling points. (b) Final centre locations.

© 1998 EURASIP

The method has been applied successfully to the location of well separated cereal grains, which can be modelled as ellipses with 2:1 aspect ratio, ready for scrutiny to assess damage, varietal purity or other relevant parameters.

10.5 Short case studies of grain and nut inspection

In this section we present a number of short case studies to give some impression of the range of subject matter covered by workers in this area. The first few of these relate to wheat inspection and grading, and these are followed by studies relating to maize, soybean, pistachio nuts and peanuts. Next, there are three case studies which illustrate use of other modalities—specifically, X-ray and NIR imaging—also for wheat and pistachio nuts. Finally, there is a short case study on the detection of bran in flour, which is in a class of its own but seems to be best placed within the present chapter.

10.5.1 Kernel positioning for grain measurement (Casady and Paulsen 1989)

Research has shown a correlation between visually detectable stress cracks on corn kernels and later breakage. To automate the task of grain inspection so as to locate such cracks reliably, or to measure grains accurately, rapid presentation of orientated grain samples is required. This paper describes a vibratory bowl feeder coupled with a chute and a cam-operated orientating device capable of feeding sample grains to the image processing platform with an accuracy such that major and minor diameters could be measured within 0.1 mm. Sensors were able to detect malfunctions and provide suitable corrective measures: error-free operation occurred in 99% of cases.

10.5.2 A dedicated wheat grain image analyser (Keefe 1992)

This research was aimed at the characterization of wheat grains of a variety of species so that they could be recognized rigorously at any site. A system was devised for extracting relevant visual features from wheat grains. Grains of 20 varieties of wheat were then fed to the system which took 33 measurements of each grain, and then computed a further 36 derived parameters. A nearest neighbour classifier was trained on 50 grains of each species using the 69 parameters, and a further 500 grains of each species were tested using the classifier. This led to a 'confusion matrix'—a 20×20 table of cases showing which species were interpreted as other species. A confusion matrix would normally be expected to be approximately symmetric, as it reflects similarity between pairs of species.

For the same reason, the largest components of such a matrix are along the leading diagonal. In this case many off-diagonal elements of the matrix were far from zero, and it was concluded that there was an overlap of appearance between grain varieties, which was made worse when grains became damaged or deformed by the shrivelling that often occurs during ripening. This meant that it was impossible to identify every grain uniquely. Nevertheless, it seems likely that it will be possible to characterize a load of grain so that it will be clear when a single variety delivery contract has been breached. Thus millers and grain merchants will have a useful tool for verifying the varietal purity of a load of grain. It was anticipated that the rate of sample throughput by such a system would eventually be limited by the speed of grain presentation rather than the speed of image processing: in the research reported, each of these operations took 2.5 minutes to process 50 grains. The grains were viewed from the side when they were lying crease down, and the resolution used was 33 pixels per mm.

10.5.3 Discrimination between wheat grain varieties and classes (Zayas and Steele 1990)

This research was aimed at discrimination between (1) visibly similar wheat varieties; (2) hard and soft wheat; (3) spring and winter wheat; (4) foreign material and wheat; (5) whole and broken kernels; (6) normal and mould damaged samples; and (7) milled fractions of hard and soft wheat. Top views of crease-down kernels were imaged and 10 features including width, length, aspect ratio, vertex angle and normalized grey-level were extracted. Multivariate discriminant analysis was carried out to permit discrimination of types 1, 2 and 3 listed above, the classification being about 80% correct. Multivariate analysis and comparison with a wheat prototype were used for discrimination of type 4, with very high success rates by the latter method. Type 5 discrimination required twelve further morphometrical measurements, including perimeter, convex perimeter, and compactness, giving around 99% correct classification. Type 6 and 7 discriminations were assessed by colour analysis and by image texture analysis respectively. The studies demonstrated significant potential for assessment of grain quality by image processing.

10.5.4 Classification of wheat kernels (Thomson and Pomeranz 1991)

This study used a laser scanning system to gather 3D information about wheat grain shapes. The laser beam was scanned from above, and the sensing photodiode was placed first one side and then the other, in order to ensure viewing the whole top surface of each crease-down grain. Dual measurements of height and intensity at each pixel location on the grain were obtained. 63 features or their combinations and ratios were extracted from the dual images and tested. Fourteen major features were identified—of which nine resulted from height measurements—and subjected to discriminant analysis. This permitted correct classification between two species of wheat kernels in ~93% of cases. Features that measure deformations at the germ end of the kernel were crucial to the classification capability. Height information provided intrinsically better discrimination than intensity information: indeed, it appeared that intensity features were more related to the environmental history of the grains than to their shapes. Finally, the system was also able to distinguish between sprouted and non-sprouted grains with ~86% success. The scale of the problem is illustrated by Figure 10.11.

10.5.5 Detection of maize and soybean quality factors (Paulsen et al. 1989)

The principal axes of arbitrarily orientated maize kernels were located by diagonalizing the covariance matrix of thresholded images after the centroid had been found. This permitted length and a width measure of the kernels to be determined: the area was found by pixel counting. Length estimation using this technique was within 0.1 mm (2–3 pixels). Soybeans were inspected for fungal damage by estimating the intensity of reflected light and textural intensity variations, as indicated by first derivatives of intensity and a regional 'change factor'. Each of the three parameters had to pass a threshold test to pass a soybean as of acceptable quality. Simple R,G,B colour tests were used to distinguish between different types of fungus. Detection of fungal damage was ~98% accurate, while classification of fungi varied from 66% to 96% depending on type.

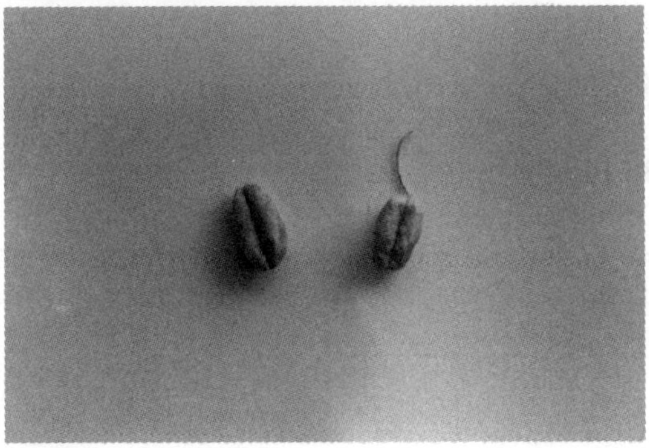

Figure 10.11 Wheat grains. This picture shows one normal grain and one which has just sprouted: in many cases the latter would not be acceptable for food processing.

10.5.6 Detection of surface defects of maize kernels (Liao et al. 1994)

This work aimed at the design of a real-time system for determination of colour and surface details of maize kernels. Maize kernel orientation was determined by measurement of second order moments of the binary object shape. Otherwise, the length, width measure and area were used as in the preceding case study. Height was also measured, from a camera positioned to obtain a side view; all HSI colour components were calculated from the R,G,B values, but the blue component was found to give the most accurate results for separating white and yellow maize. 32 Fourier boundary descriptors were computed and fed to a backpropagation neural network. The latter was able to detect broken kernels with an accuracy of around 95%. Accuracies of length, width and height dimensions were around 2–3% (0.2 mm).

10.5.7 Detection of early split pistachio nuts (Pearson and Slaughter 1996)

The shells of pistachio nuts normally split from the hull about one month before harvest, but occasionally the split causes the hull itself to rupture. The resulting 'early split' of the hull is susceptible to infestation by moulds and insects, the former of which sometimes generates aflatoxin—a known carcinogen. This research aimed to detect early split by computer vision. The method used was to form an intensity profile of each nut along pixel rows parallel to the minor axis of the nut shape (the nuts are approximately elliptical in shape). The number of rows with a characteristic shape of minimum in the intensity profile was counted and used together with the area parameter to form a linear classifier. This proved capable of distinguishing early split nuts, the effective misclassification rate being just 1%.

10.5.8 Detection of stained pistachio nuts (Pearson 1996)

Stained pistachio nuts are less attractive to consumers and have higher incidence of aflatoxin contamination. This research aimed to detect stains by image processing. The nuts were made to fall down a chute at about 45° to the horizontal, after which they passed through a cell where they were illuminated from three directions and viewed by three line-scan cameras in a ring, so that the whole surface could be inspected. Counts of pixels meeting particular intensity variation characteristics—high slope, moderate slope and low flat intensity—were recorded and used for discriminant analysis. Linear classifiers were found to be suitable, the misclassification rate being around 15%, most of which was probably attributable to the 13% variation in speed of the nuts as they passed through the observation cell.

10.5.9 Automatic grading of peanuts (Dowell 1992)

This work was aimed at automatic grading of peanuts by machine vision in combination with other methods. Kernel size was assessed using 2D monochrome imaging and finding the maximum diameter of the minor axis of the kernel: this was achieved with an accuracy of ±0.025 mm. Kernel damage was determined using grey-level texture analysis, though additional relevant information was provided by a colorimeter employing the HSI representation. An expert system was used to assess the outputs of the various sensing systems employed to assess the kernels.

10.5.10 X-ray detection of weevils in wheat (Keagy and Schatzki 1993)

This work aimed to detect granary weevils and maize weevils in wheat grains. The early stages of insect growth within wheat grains are not easily seen in X-ray images, even by human operators: typically the latter achieve 50% recognition. However, it proved possible to initiate automatic recognition of the structural damage caused by insects by looking for intersections between internal edges and the crease lines of the grains. First a Laplacian operator was applied to enhance the lines. Then L-shaped and V-shaped binary convolution masks were applied to detect the intersections; those adjacent to the grain background were eliminated by application of a Sobel operator. At least eight L-shaped and eight V-shaped masks of different orientations were required in order to limit the numbers of false positives; the masks also had to be at least 5 × 5 or 7 × 7 pixels in size for the same reason—as in the following basic mask patterns:

1						1				1
1						1			1	
1						1		1		
1						1	1			
1	1	1	1	1		1				

In spite of this intrinsically reasonable strategy for insect location, it was found that "machine recognition of an insect is a chance event". In addition, the false positive problem was a serious consideration: in practice the false positive rate would have to be limited to 0.5% or further verification of infestation would be required before rejection of a consignment of grain.

10.5.11 X-ray detection of insects in pistachio nuts (Casasent et al. 1997)

This research aimed at a reliable means for detecting insects in pistachio nuts using X-ray imaging. First, a watershed algorithm (Chapter 12) was used to segment touching or overlapping nuts from each other. Next, a set of basic images (including the original image) was derived by edge detection and detection of curvature in the original and the edge images. Any highlights at the grain boundaries of these images were eliminated as irrelevant to internal infestation, and various thresholding and normalization operations were applied. At this point the mean, variance, skewness and kurtosis of the images were computed, and the resulting 16 parameters analysed to determine the best ones to use for insect detection. Use of five features—all four variances, plus kurtosis of the original image—were found to give the most statistically significant result. A quadratic statistical classifier was tested and found not to give as good classification rates as a 3-layer piecewise quadratic neural network which was capable of handling several disjoint quadratic decision surfaces per class. However, as in the previous case study, there were distinct problems with false positive rates if high classification rates for detecting infestation were to be attained. Taking a false positive rate of 2% as permissible, the probability of correct classification for infested nuts was only ~66%. At this point it should be noted that the incidence of infestation was around 5%. Thus 66% of these were located, so 34% were not detected, amounting to just 1.6% of non-detected infested nuts in a typical consignment. This appeared to be a reasonable tradeoff against the 2% false positive rate.

Comment. The tradeoff between false positives and false negatives is a fundamental problem in practical statistical pattern recognition. It is not generally the case that the minimum overall error rate represents the optimum solution: optimization will normally be better calculated in terms of minimum losses or costs, which may mean that in any given application, the *ratio* of numbers of false positives to false negatives should be fixed in advance. However, in general the problem is tackled in terms of the 'receiver operator characteristic' (ROC), which is a plot of the true positive versus the false positive rate. The problem is well aired in the related work by Keagy *et al.* (1995, 1996) and will not be discussed further here.

10.5.12 Detection of insects inside wheat kernels by NIR imaging (Ridgway and Chambers 1998)

This paper studies the potential for detecting insects inside wheat kernels by NIR imaging. Careful visual analysis of wheat kernels imaged at NIR wavelengths showed that it should be possible to detect insects by this means. However, success may depend on imaging at two wavelengths with subtraction to sufficiently enhance the weak signals from the insects. Earlier formative work on this topic is described in Chambers and Ridgway (1996).

10.5.13 On-line measurement of bran in flour (Whitworth et al. 1997)

This work is aimed at the assessment of the quantity of bran in white flour. Specks of bran are individually identified within images of flour as it passes along a flour spout. The bran is identified by its darker colour: both the total bran content and the number of specks of bran are measured from the images. The system has been commercialized and uses a sampling rate of 3 per minute to permit the analysis to be carried out on a PC. The results are accurate to about 0.1%—a degree of accuracy that is attained with the aid of careful calibration against test cards. This advance is typical of those that need to be achieved if the industry is to overcome the problems of variability between human inspectors.

10.6 Concluding remarks

Almost the whole of this chapter has been devoted to case studies, three of which were substantial, and a further thirteen of which were short as they were intended to indicate the variety and richness of the subject area. In principle, mere presentation of case studies would not represent a good balance. However, in this chapter at least, such a criticism is not justified: the major case studies presented quite new or topical image processing techniques in the right context; in any case, the chapter should not be seen in isolation, but rather in conjunction with the earlier work of Part 1 which presented the basic concepts and methodology of the subject.

Nevertheless, case studies are intended to take us forward, to see both what areas people are working on and what techniques are available—and also what their limitations and advantages are. It is one thing to present idealizations of the image processing and quite another to see whether they match up to the stark realities of applications in the present day. We have already heard about 'nasty realities' in image processing: in particular, it is worth recalling that algorithms must be reliable, robust and sufficiently fast to make it realistic to use them in practical food applications. The case study in Section 10.4.2 gave some good indications of the ultimate speeds of image processing algorithms, though, clearly, use of hardware accelerators of various sorts can improve speeds further, if at some additional cost.

But what of the topics covered in this chapter? First, the foods mentioned included wheat and other cereal grains, maize, and also various sorts of nuts. Contaminants, fungi and insect infestations were also considered. Then there were the imaging modalities for inspecting the foods: these covered visible light, X-ray and NIR sensing. In addition, colour imaging was found to be useful for separating white and yellow maize, and for detecting moulds on grain.

The techniques that were found useful included convolutions, morphological processing, and a variety of simple shape measures. Equally important were considerations of false positive detection rates and the receiver operator characteristic method for optimizing the balance between true and false positive rates. Overall, inspection and grading of grain and nuts are of some importance in the food industry, and improving classifier discrimination performance is a challenging task for researchers.

Finally, lest it be thought that a single chapter can fully describe all the work that has been completed in this area, the reader's attention is drawn to the recent work of Majumdar *et al.* (1997) on cereal grain classification, and a number of papers relating to seed identification and certification: Churchill *et al.* (1992); Peterson and Krutz (1992); Shatadal *et al.* (1995); Chtioui *et al.* (1996). In fact, all these papers involve image processing techniques that have been covered in some depth within the confines of this monograph.

Chapter 11

X-ray inspection

11.1 Introduction

This chapter discusses methods of X-ray inspection. The particular value of X-ray inspection has already been pointed out in earlier chapters. Specifically, it is useful because it permits the final packaged products to be inspected just before they are sent for distribution to customers: this makes the whole inspection process much more rigorous, because contaminants can conceivably be added right up to the time the products leave the factory. Inclusion of final X-ray inspection is therefore a good indication of due diligence by the manufacturer.

The chapter looks first at the process of X-ray image formation. It then presents several case studies of the application of X-ray inspection in the food industry. The first three of these describe the analysis of X-ray images of frozen food packs. The problem is complex because no single analysis technique is capable of recognizing all possible contaminants, and three case studies are needed to cover the main issues. The first case study follows the fairly obvious path of locating foreign objects by thresholding: this should be effective for the location of pieces of metal with high X-ray scattering cross sections, and if successfully implemented should also be capable of locating several other hard contaminants. The second case study employs other techniques which will be useful if the intensity patterns of the substrate food are textured, so that more complex processing is required: in such cases thresholding is not expected to be successful, even though it may still be able to locate specific hard contaminants. The third case study considers the hardware needed to implement the frozen food pack inspection algorithms in real time. A final set of short case studies follows, and includes some which are concerned with the inspection of metal and glass containers and the interior regions of fruit and vegetables.

11.2 X-ray image acquisition

The value of X-ray inspection has already been pointed out. In this section we consider briefly the technology that is required for this purpose.

Commercial X-ray inspection systems are quite expensive and typically cost over £40,000, i.e. about an order of magnitude more than visual inspection systems. The significantly greater relative cost stems largely from the substantial safety measures needed with X-ray systems. Not only is a considerable amount of shielding needed to protect human operators, but this necessitates a conveyor being built into the system and the machine being marketed as a whole. In addition, the machine requires electronic interlocks and calibration facilities, together with relatively new types of sensor which are rather more expensive than their optical counterparts. The X-ray photons from a special tube in which high energy electrons are decelerated by a carefully chosen metal target are passed through the substance to be investigated. They then pass into a layer of scintillating material where they generate visible light photons which are detected by a linear photodiode array. The array extends over a substantial distance across the width of the conveyor on which the food substances are placed, but in other respects it resembles the linear photodiode array of a CCD line-scan camera (though the resolution is likely to be rather lower, closer to 512 than to 4000 pixels). It should be remarked that such systems also have quite large maintenance costs, as the X-ray tubes have to be replaced regularly; in addition, the sensors deteriorate with time because of radiation damage, and also have to be replaced periodically. Ongoing losses in sensitivity also demand frequent (e.g. daily) recalibration of the system.

The line-scan camera system grabs images as the products move along the conveyor. The penetration of the X-rays can be changed by adjusting the voltage at which the X-ray tube operates. Similarly, the intensity of the X-ray beam can be changed by adjusting the electron current in the X-ray tube. Reasonable beam intensities can be obtained at convenient small currents in the range 3–10 mA, and these are well within the permitted radiation levels for foodstuffs. The adjustment of the X-ray tube voltages is more problematic, as highly penetrating beams will pass right through the products with negligible attenuation, and little information will be obtainable. Clearly, there is an intermediate level of penetration at which contrast is optimal for any given object, but in reality the level of penetration will need to be targeted at a specific thickness of a specific type of material.

The situation is complicated by the complex scattering of X-rays as they pass through thick substances. First, X-rays will reach a particular layer of material after passing through higher layers, and as a result of the scattering will eventually reach the sensors via numerous indirect paths. Second, during its path to the sensor, the X-ray beam will be hardened as more and more of the softer rays are absorbed by the material. The simplest model of the process involves considering only the softer X-rays: these can be assumed to be subject to exponential decay as they pass through a number of layers of a homogeneous substance of thickness z:

$$I = I_0 \exp\left[-\int_0^z \mu(z) dz \right] = I_0 \, e^{-\mu z} \tag{11.1}$$

where μ depends on the type of material and the penetrating power of the X-ray beam. It turns out that μ varies inversely with the energy E of the photons. Detailed analysis shows that, under certain simple assumptions (Davies 1997e), the contrast of the X-ray beam is optimized for a value of $\int \mu \, dz$ (the generalized beam path length) proportional to E.

Another relevant point is that because the attenuation of the beam is exponential in the distance travelled by the beam through the material, the image data embodies a type of distortion. Thus it will in many cases be best to linearize the data by taking the logarithm of the input signal, before proceeding to use it for making physical measurements on products.

11.3 Case study: reliable thresholding of X-ray images

The case study outlined in this section involves the inspection of bags of frozen vegetables, such as peas and sweetcorn. The bags are inspected at the end of the product line, after sealing, since this provides the most rigorous test of the final quality of the product. However, since the bags are opaque and much thicker than a single layer of vegetables, X-ray inspection is required if any foreign objects such as pieces of metal are to be located.

In this case study it is assumed that the X-ray images obtained from the substrate food are sufficiently uniform that some contaminants can be located by thresholding (Hannah *et al.* 1995). However, there is a problem in reliably estimating the threshold intensity values to be used. This problem is especially difficult if the contaminants are small. For instance they may be metal staples or splinters of glass; in some cases the widths of the metal pieces to be detected may be of sub-pixel width, so that only very few pixels may give a substantial signal. Nevertheless, thresholding should lead to detection if the threshold level is properly selected. The question now arises of how this may be achieved.

The standard approach to thresholding outlined in Chapter 2 involved finding the neck of the global image intensity histogram. However, this is impracticable when the dark peak of the histogram is minuscule in size, as it will then be hidden amongst the noise in the histogram and it will not be possible to extract it with the usual algorithms.

A good many investigators have tackled the thresholding problem (e.g. Otsu 1979; Kittler *et al.* 1985; Sahoo *et al.* 1988; Abutaleb 1989): amongst the most well-known approaches are the variance-based methods. In these methods, the image intensity histogram is analysed to find where it can best be partitioned to optimize criteria based on ratios of the within–class, between–class and total variance. The simplest approach is to calculate the between–class variance, as described below (Otsu 1979).

11.3.1 Variance-based thresholding measures

First we assume that the image has a grey-scale resolution of L grey levels. The number of pixels with grey level i is written as n_i, so the total number of pixels in the image is $N = n_1 + n_2 + \cdots + n_L$. Thus the probability of a pixel having grey level i is:

$$p_i = n_i / N \tag{11.2}$$

where:

$$p_i \geq 0 \qquad \sum_{i=1}^{L} p_i = 1 \tag{11.3}$$

For ranges of intensities up to and above the threshold value k, we can now calculate the between–class variance σ_B^2 and the total variance σ_T^2:

$$\sigma_B^2 = \pi_0(\mu_0 - \mu_T)^2 + \pi_1(\mu_1 - \mu_T)^2 \tag{11.4}$$

$$\sigma_T^2 = \sum_{i=1}^{L} (i - \mu_T)^2 p_i \tag{11.5}$$

where:

$$\pi_0 = \sum_{i=1}^{k} p_i \qquad \pi_1 = \sum_{i=k+1}^{L} p_i = 1 - \pi_0 \tag{11.6}$$

$$\mu_0 = \sum_{i=1}^{k} ip_i / \pi_0 \qquad \mu_1 = \sum_{i=k+1}^{L} ip_i / \pi_1 \qquad \mu_T = \sum_{i=1}^{L} ip_i \tag{11.7}$$

Making use of the latter definitions, the formula for the between–class variance can be simplified to:

$$\sigma_B^2 = \pi_0 \pi_1 (\mu_1 - \mu_0)^2 \tag{11.8}$$

For a single threshold the criterion to be maximized is the ratio of the between–class variance to the total variance:

$$\eta = \sigma_B^2 / \sigma_T^2 \tag{11.9}$$

However, the total variance is constant for a given image histogram, so maximizing η simplifies to maximizing the between–class variance.

The method can readily be extended to the dual threshold case $1 \le k_1 \le k_2 \le L$, where the resultant classes C_0, C_1 and C_2, have respective grey-level ranges of $[1,\ldots,k_1]$, $[k_1+1,\ldots,k_2]$ and $[k_2+1,\ldots,L]$.

Unfortunately, it was found that these two methods did not match the requirements for the X-ray contaminant detection task, because the small sizes of the foreign objects being sought prevented their intensity distributions from being discerned against the noise of the histogram (Hannah *et al.* 1995). Thus a more sensitive method was required. This led to tests on entropy-based thresholding measures, which have also become important over the past decade or so (Pun 1980; Kapur *et al.* 1985; Abutaleb 1989; Brink 1992).

11.3.2 Entropy-based thresholding measures

Entropy measures of thresholding are based on the concept of entropy. The entropy statistic is high if a variable is well distributed over the available range, and low if it is well ordered and narrowly distributed: specifically, entropy is a measure of disorder, and is zero for a perfectly ordered system. The concept of entropy thresholding is to threshold at an intensity for which the sum of the entropies of the two intensity probability distributions thereby separated is maximized. The reason for this is to obtain the greatest reduction in entropy—i.e. the greatest increase in order—by applying the threshold: in other words, the most appropriate threshold level is the one which imposes the greatest order on the system, and thus leads to the most meaningful result.

To proceed, the intensity probability distribution is again divided into two classes—those with grey levels up to the threshold value k, and those with grey levels above k (Kapur *et al.* 1985). This leads to two probability distributions A and B:

$$\text{A:} \qquad \frac{p_1}{P_k}, \frac{p_2}{P_k}, \ldots, \frac{p_k}{P_k} \tag{11.10}$$

$$\text{B:} \qquad \frac{p_{k+1}}{1-P_k}, \frac{p_{k+2}}{1-P_k}, \ldots, \frac{p_L}{1-P_k} \tag{11.11}$$

where

$$P_k = \sum_{i=1}^{k} p_i \qquad 1-P_k = \sum_{i=k+1}^{L} p_i \tag{11.12}$$

The entropies for each class are given by:

$$H(A) = -\sum_{i=1}^{k} \frac{p_i}{P_k} \ln \frac{p_i}{P_k} \tag{11.13}$$

$$H(B) = -\sum_{i=k+1}^{L} \frac{p_i}{1-P_k} \ln \frac{p_i}{1-P_k} \tag{11.14}$$

and the total entropy is:

$$\mathscr{H}(k) = H(A) + H(B) \tag{11.15}$$

Substitution leads to the final formula:

$$\mathscr{H}(k) = \ln\left(\sum_{i=1}^{k} p_i\right) + \ln\left(\sum_{i=k+1}^{L} p_i\right) - \frac{\sum_{i=1}^{k} p_i \ln p_i}{\sum_{i=1}^{k} p_i} - \frac{\sum_{i=k+1}^{L} p_i \ln p_i}{\sum_{i=k+1}^{L} p_i} \tag{11.16}$$

and it is this parameter which has to be maximized.

Again, this approach is straightforwardly extended to dual thresholds, but we shall not go into the details here (Kapur *et al.* 1985).

11.3.3 A modified entropy measure

In the particular X-ray inspection task, it was necessary to employ two thresholding levels. One threshold related to distinguishing the darker from the lighter parts of the food substrate. A second threshold was therefore required to locate the contaminants. Thus the dual thresholding approach was the obvious choice. Clearly, it would be simpler if a single threshold could be used to locate the contaminants, as it is not necessary to separate the darker and lighter parts of the food substrate. However, in practice the task is dominated by the weight of the data which always points first to the rather irrelevant intra-substrate threshold. (For some less textured products, the first threshold relates to distinguishing the substrate from the conveyor—though this will not often be necessary.)

When applying dual threshold entropy thresholding, two main problems are found to arise: two thresholds are always obtained, irrespective of their significance, and in particular, irrespective of whether any contaminants are present or not; the method is significantly more computation intensive than the single threshold method as it involves search over a 2D manifold. These considerations led to development of the alternative strategy shown in Table 11.1 (Hannah *et al.* 1995). This strategy has the following advantages:

- It does not force two thresholds to be applied unless this is appropriate.
- It is faster to compute than the dual threshold approach.
- It can be made sensitive to small foreign objects which are often missed by alternative thresholding methods.

The last property is particularly useful since the foreign objects that can occur in food packs can be as small as a single pixel in size. The sensitivity of the method to the object size is set by determining the sizes of the peaks sought by the method. A simple peak detection scheme that depends on two factors—the width of the peak, and the height of the peak compared with adjacent values—was devised. Typically, a peak is detected if the entropy measure decreases for two subsequent grey levels, or if the entropy measure falls to 0.96 of the preceding value. By appropriate manipulation of these factors the method can be made to detect smaller foreign objects, while also producing a lower percentage of false rejects (rejecting bags containing no foreign objects).

Overall, the revised strategy seems to outperform the other strategies tested, including the variance-based methods and the basic single and dual entropy thresholding methods. The ability to set the sensitivity in order to detect foreign objects of different sizes is a marked advantage over the other methods, as is the capability for not forcing two thresholds

Table 11.1 Alternative strategy for dual threshold selection

1. Determine the major peak (k_1) using the single threshold method.
2. Search below k_1 for the lowest alternative peak.
3. Search above k_1 for the highest alternative peak.
4. If low and high alternative peaks both exist, determine which is the major peak (k_2).
5. If only one alternative peak exists assign it to k_2.
6. Threshold the image at k_1 if appropriate.
7. If an alternative peak has been found, threshold the image at k_2.

inappropriately. Finally, the improved performance relative to the dual threshold approach seems to be at least partly due to the better statistics resulting from working in a 1D rather than a 2D manifold: this notion is supported by the fact that the two methods often lead to substantially different threshold levels.

Next, we illustrate the operation of the revised strategy. The histogram of the image in Figure 11.1(a) is shown in Figure 11.1(b). The corresponding entropy measure is shown in Figure 11.1(c). The histogram in Figure 11.1(b) exhibits a sharp peak at the high grey level values associated with the background conveyor belt. There is also a broader peak produced by the frozen food, in the mid to lower grey level range. The peak associated with the small dark foreign object does not show at this scale. However, the entropy measure (Figure 11.1(c)) shows both a major, broad peak associated with the frozen food grey levels and also a second sharper peak at lower grey levels: this is the peak associated with the foreign object. In view of the size of the second peak it can be said to contain important information about the image and it represents a plausible alternative level where the image can be thresholded to detect the foreign object. The revised method searches the entropy function for alternative peaks of this sort, and then thresholds at the single or dual thresholds as appropriate.

Figure 11.1(d) shows the results of using this method to determine the thresholds for the image in Figure 11.1(a). As can be seen from Figure 11.1(d) the foreign object has been detected along with some background noise points. In fact, the dual threshold method was not able to detect the foreign object present in the image shown in Figure 11.1(a).

Finally, one danger associated with increased sensitivity in the detection of small foreign objects ought to be mentioned: this is a slightly increased likelihood of false positive detection. On the other hand, it will generally be better for the food industry to err on the side of over-caution when setting the sensitivity of a foreign object detection method than to let through a large number of products containing foreign objects.

11.4 Case study: inspection of frozen food packs

The case study outlined in this section also relates to the inspection of bags of frozen vegetables (Patel *et al.* 1996). It will be apparent from the previous discussion that X-ray inspection systems that rely on simple thresholding have limited sensitivity. In particular, they tend to be good at detecting pieces of metal, but poor at locating so-called 'soft' contaminants such as wood, plastic and rubber; in addition, they are often unable to detect small stones even though these are commonly classed as 'hard' contaminants. One basic problem is the high level of intensity variation in the X-ray image of the substrate vegetable matter: the fact that several layers of vegetables

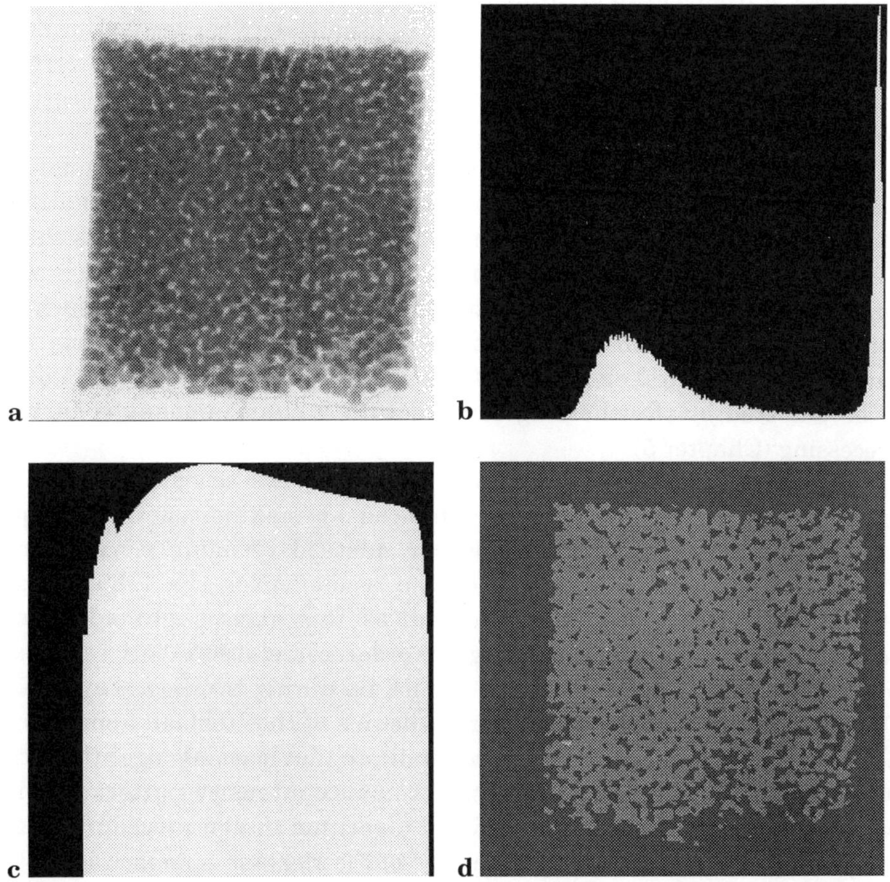

Figure 11.1 Entropy thresholding. (a) X-ray image of a bag of sweetcorn containing a piece of glass approximately a third of the way up the left hand edge of the bag. (b) Intensity histogram of (a). (c) Result of entropy calculation. (d) Final thresholded version of (a), using the procedure described in the text.

contribute to the same image means that the latter appears highly textured, and it is rather ineffective to apply simple thresholding.

11.4.1 General strategy of the system

In the application considered here, no assumptions can be made about the individual foreign objects that might occur, so the usual algorithms for locating defects cannot be used. In particular, shape analysis and simple measures of intensity are mostly inappropriate, and thus it is necessary to recognize foreign objects by the fact that they disrupt the normal (textural) intensity pattern of the substrate. For example, it might have been thought that a set of feedforward artificial neural networks (ANNs), each adapted to detect a particular foreign object, would be useful (see Chapter 7). However, so many types of foreign object with so many possible shapes and sizes can in principle occur that this is not a viable approach except for certain crucial contaminants.

To solve this problem Laws' (1979, 1980) approach to texture analysis was adopted rather than the widely used grey-scale co-occurrence matrices or the other techniques referred to in Chapter 5. The reasons for this choice were (a) ease of setting up and (b) the fact that Laws' approach is well adapted to hardware implementation as it employs small neighbourhood convolutions to obtain a set of processed images. In this application, summing the 'textural energies' in these images permits any foreign objects to be detected by thresholding coupled with a minor amount of further processing (Chapter 5).

Following Ade's (1983) modification of the Laws' schema (Chapter 5), it was found that sensitivity is enhanced by making use of principal components analysis (PCA). However, instead of using conventional diagonalization procedures, the Hebbian type of ANN (Oja 1989; Sanger 1989; Patel *et al.* 1994) was adopted, as this gives slight additional sensitivity—probably because higher order correlations between pixel intensities which the ANN is able to pick up during training (Taylor and Coombes 1993). Space does not permit details of the Hebbian approach to be described in detail here, but it does indicate that learning algorithms can be particularly useful for processing complex intensity patterns which would lead to problems if conventional algorithm design procedures were insisted upon. Another major advantage of the Hebbian approach is that it permits PCA to be applied without the huge computational load that would be expected when dealing with large matrices; in addition, the Hebbian approach can be implemented iteratively in such a way that additional training patterns can be applied in an ongoing fashion, permitting continuous adaptation to the state of the product line.

One more detail will be of interest. By adopting a statistical pattern recognition approach, it is possible to classify the images into three regions—background region, foodbag region, and any foreign object regions. Thus it is unnecessary to have a preliminary stage of bag location—with the result that the whole inspection algorithm becomes significantly more efficient. It appears that this approach will be widely applicable when the objects being inspected do not demand prior structural analysis: indeed, if in this application only loose product had appeared on conveyors, it would have been unnecessary to have a third classification category, except perhaps to report on any temporary lack of product.

11.4.2 Further details of the system

The considerations mentioned above lead to a system design of the form shown in Figure 11.2. In particular, the initial acquisition stage is followed by a preprocessing stage, a feature extraction stage and a decision stage. Note that the Hebbian training paths are not shown in this figure, which just includes the data paths for normal testing of input images.

The main component of Figure 11.2 which has not been covered by the earlier discussion is the preprocessing stage. In fact, this has several components. The first is the log transform: this is a vital part of the overall system, as it compensates for the non-linearity of the image acquisition process, thereby making the occupation levels of the grey-levels more uniform and the subsequent processing more reliable. It should also be mentioned that there is further justification for this procedure as it provides an element of noise whitening and should therefore lead to improved sensitivity of any subsequent detection process which approximates to matched filtering.

Rank-order filtering provides further capabilities for preprocessing. In particular, local intensity minimization operations have been found valuable for expanding small dark foreign objects in order to make them more easily discernible. In some cases the same operation has also been found useful for enhancing the contrast between soft contaminants and the food substrate.

Lastly, thresholding is incorporated into the predominantly texture analysis scheme, both to provide the capability for locating contaminants directly (as described in the previous case study) and as the final decision making stage of the texture analysis process.

There is no space here to give full details of the tests carried out to

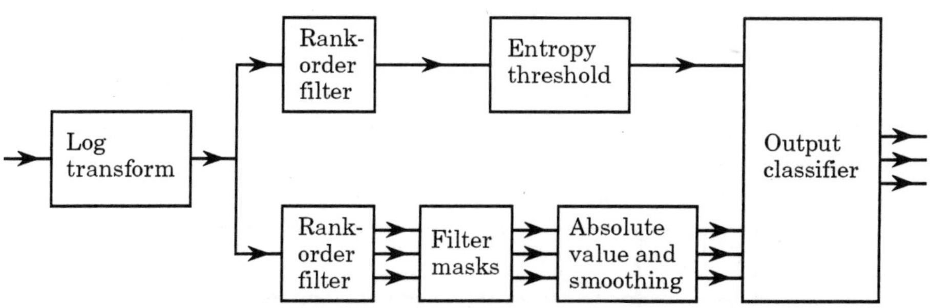

Figure 11.2 Foreign object detection system. The input image comes in on the left, and the output classification (which is not an image) emerges on the right. Entropy threshold and texture energy images are also available and may be recorded to demonstrate 'due diligence'.

apply the above techniques to X-ray inspection of frozen food packs (Patel *et al.* 1996). Instead we summarize the situation. It was found to be both effective and computationally efficient to use Laws' masks of size 3 × 3 to form the microfeatures, and, following absolute value determination, to use smoothing masks of size 5 × 5 to obtain the texture energy macrofeatures. The tests were made with 1 lb. bags of frozen sweetcorn kernels into which foreign objects of various shapes, sizes and origins were inserted: specifically, foreign objects consisting of small pieces of glass, metal and stone and larger pieces of plastic, rubber and wood were used for this purpose (Figure 11.3).

Table 11.2 compares the recognition rates achieved by human operators and by the final system. The 'expert inspector' was disadvantaged in having to view the images unaided, while the 'human operator' was allowed to progressively threshold the images manually in order to fine-tune the system. The *minimum* of the two human results could be taken as indicative of the level of reliability and consistency typically achievable by human inspectors, while the *maximum* of the two could be taken as indicative of the ultimate level of sensitivity that should be aimed for in an automatic inspection system. It is gratifying that, except in the case of wood, the final inspection system exceeded these expectations by a small margin. Poor performance for wood is understandable as its characteristics are not at all dissimilar to that of sweetcorn, which is also vegetable matter; in addition, it presents a difficult case because of its low X-ray absorption cross section.

Finally, it is important to note that the types of training procedure employed in the algorithms have been found to make them highly adaptable to inspection of a wide variety of products, including not only vegetables, but also 'stir-fry' (which contains chunks of meat), briquettes of breakfast cereal, and chocolate bars of various types (e.g. plain chocolate and chocolate covered wafers) (Patel *et al.* 1995).

Table 11.2 Recognition rates achievable by human operators and by the inspection system

foreign object	expert inspector	human operator	inspection system
glass	50	80	80
metal	84	92	96
stone	100	100	100
rubber	64	40	65
wood	32	36	20

11.5 Case study: design of hardware for inspection of frozen food packs

This case study is concerned with the design of a real-time system for the detection of contaminants in images of food products where the intensity variations are highly textured—as described in the previous case study.

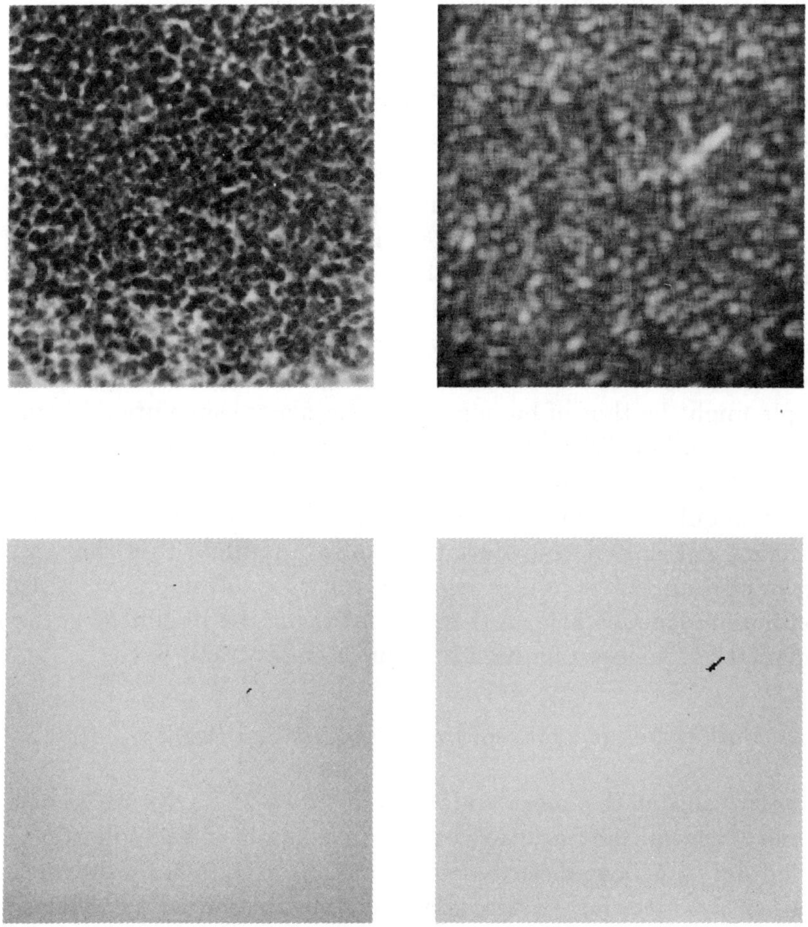

a b c d

Figure 11.3 Foreign object detection using texture analysis. (a) shows the original X-ray image of a packet of frozen sweetcorn. (b) shows an image in which any foreign objects (here a splinter of glass) have been enhanced by texture analysis. (c) and (d) show the respective thresholded images. Notice that false alarms are starting to arise in (c), whereas (d) gives much increased confidence in the detection of foreign objects.

Processing the images in real time presents a very significant problem, not only in throughput but also regarding system cost and certain other issues. The case study analyses the problem and concludes that the most competitive type of real-time solution is software based, thereby aiding system adaptability and flexibility. The case study also describes a particular software-based hardware system that can inspect the products in real time (Davies *et al.* 1995).

11.5.1 Real-time processing issues

The problem of optimal hardware implementation has been tackled by Davies and Johnstone (1986, 1989) in the context of food product inspection. While this approach provides a methodology which is able to answer a number of vital questions about the optimal form of any real-time implementation, there are still some issues which it does not address. An important one is that of system flexibility. For example, if some optimal dedicated hardware solution is implemented in a given application, and the situation changes and the system has to be modified appropriately, the problem of adaptation to the new situation is not tackled. In fact it is evident that a software solution of some sort will be far more flexible. One example might be that of buying one of the many convolution boards that are on the market. While these are in general highly cost-effective, some sort of bit-slice, or transputer, or digital signal processor (DSP) chip solution might prove an even better buy as it could be used flexibly to implement not only a convolution but also any algorithm that could be programmed on the computer used for software development. Naturally, the situation depends also on the relevant costs, the likelihood of the need for adaptation, the need for high-level functionality, and so on.

11.5.2 Main contenders for real-time inspection hardware

Currently, one of the most widely used types of hardware in use is a pipeline of specialized hardware cards. An example of its application in the food industry has been described by Penman *et al.* (1992) for the inspection of canned products for foreign material—see Section 11.6.8. This system comprises a Motorola 68020 single-board computer and a range of Datacube MaxVideo image processing modules. The Datacube MaxVideo modules are pipeline processors which can perform a wide range of processing operations at video rates. The overall processing speed may be increased either by cascading more modules or by operating several pipelines concurrently. The Motorola 68020 controls the entire image processing operation including the Datacube pipeline modules that are accessible over the VME bus.

However, it appeared that this particular methodology would not be sufficiently flexible for the application considered here, and this led to the adoption of a software approach based on an array of single-stream computers: in fact, it seemed that the Centurion machine (see below) would be particularly well adapted to the task. At this stage it had to be demonstrated whether the Centurion (or other similar system) is able to operate sufficiently rapidly on the most computation-intensive part of the task, which in this texture analysis application is the low-level convolution-based analysis: note particularly that this analysis is applied many times to all the pixels in the input and derived images, whereas subsequent processing (e.g. to identify, locate and count any foreign objects in foodstuffs) is only applied to selected areas of the images or to abstract data extracted from these images. In fact, this case study only considers the convolution aspects of the algorithms, which constitute about 80% of the processing necessary to locate the crucial contaminants.

The timing results presented in Table 11.3 show that the Centurion is easily able to cope with the real-time demands of such applications—previously thought to be impractical without the support of SIMD processors or considerable amounts of dedicated hardware. Since the Centurion implementation is software using a medium-parallel array of sequential processors, it offers far greater flexibility than either SIMD or hardwired solutions, with significant processor power available for remanent non-parallel tasks. Perhaps most important of all, the implementation cost of the system is substantially less than £10K, compared with at least £25K for a fixed function hardware-based system.

Table 11.3 Timings showing the performance of the real-time system. Timings of the system were obtained using 1–3 array boards (each containing eight 2115 DSPs), at both low (380 × 256) and high (760 × 512) resolution. For comparison, a typical foreign object detection application would require one bag of frozen vegetables to be inspected per second, this being easily achieved with a single array board operating at low resolution

no. of Array Boards	Resolution	
	380 × 256	760 × 512
1	133.8 ms	527.5 ms
2	72.2 ms	272.9 ms
3	49.3 ms	186.7 ms

11.5.3 The Centurion software-based real-time system

The availability of low-cost Digital Signal Processing (DSP) devices holds out the promise of highly cost-effective image processing systems. However, to match the demands of real-time applications, many such processors need to be coupled together.

The Centurion system (Image Inspection 1994) is based around a linear array of up to 120 Analog Devices ADSP 2115 DSP devices (Analog Devices 1990). In normal use, images are split up into vertical overlapping strips which are broadcast to the array processors via a 20 Mbyte s^{-1} backplane bus. The system also contains a master controller board which provides additional processing capability and an 80186 microprocessor which can run programs written in C and compiled using industry standard compilers. The standard software includes a large set of standard image processing operations which may be called through an internal scripting language running on the 80186 processor. Typical system configurations cost £10–15K.

11.5.3.1 Communication channels

The Centurion memory is distributed with the processors. There is no shared memory, but standard code fragments in the processors' program stores can be used both to broadcast data to all processors and to target individual processors.

Multiprocessor systems require efficient inter-processor communication channels. The Centurion offers two pathways which may operate in parallel—a 20 Mbyte s^{-1} shared backplane, which may be segmented into sub-channels connecting clusters of eight processors, and a set of full-duplex synchronous serial channels that connect adjacent array processors in a manner reminiscent of transputer channels. The serial channels include a direct memory access capability that allows block transfers to be run in the background without program intervention.

The backplane is used mainly for the transfer of entire video images between the video board and the array. Code in each processor allows the array elements to capture their portions of the data broadcast from the video board, and to transmit their segments in the correct sequence when collection is occurring. The backplane supports 16-bit data transfers at 10 MHz for an aggregate transfer rate of 20 Mbyte s^{-1}.

11.5.3.2 System components

A Centurion system must contain at least one master board. The Centurion backplane addressing mechanism limits the system to a total of sixteen boards, so the largest array that can be assembled comprises a master board and fifteen array boards giving a total of 120 processors.

The master board provides an 80186 processor fully populated with memory including a large FLASH EPROM store which is used to hold image processing programs as a sort of solid state disk drive. The master board also includes an ADSP 2101 processor which may be used as a stand-alone image processor if no array boards are present. The 80186 and 2101 processors communicate via a bidirectional first-in first-out memory which is a queue data structure implemented directly in hardware.

The array boards all contain eight 2115 processors, each with a large local memory. The processors are interconnected via a local pipelined, bidirectional bus.

11.5.3.3 Performance

The Centurion displays near-linear speed-up for the convolution type operations described in this case study (see below). The well-known Sobel edge enhancement operator can run in less than one frame time on a suitably configured system containing 8 boards (120 DSPs).

The slight falling off in performance improvement on large arrays is an indication of the relative contribution of the overlapping of strips, necessary for seamless processing of window operations. For an $N \times N$ pixel image the intrinsic width of each strip is N/p for p processors. However, for an $n \times n$ convolution window, the width of each strip becomes $N/p + 2\lfloor n/2 \rfloor$: for the 3 × 3 Sobel window, in the case where $N = 512$ and $p = 32$, the width of each strip is, at best, increased from 16 to 18 pixels, thereby increasing the execution times by ~11%.

11.5.3.4 Use of the Centurion for foreign object detection

With the array-board add-on capability used by the Centurion, there is clearly a cost/performance tradeoff. Tests with configurations of up to three array boards resulted in the whole-image execution times given in Table 11.3. These timings were taken without implementation-specific optimization of the Centurion system, and hence are slower than in a practical implementation. Nevertheless, they are quite impressive. What is more, they represent a faster than real-time solution for the application considered here. For example, in a typical factory application, in which 1 lb. bags of frozen sweetcorn would be inspected at the rate of one per second, the low-resolution (380 × 256 pixel) implementation will be adequate, and one array board will easily achieve the necessary throughput. Furthermore, the high-resolution (760 × 512 pixel) implementation will also provide a comfortable solution, even without optimization. The reserve is valuable since it permits leeway in the implementation of the more variable types of high-level processing, which, however, we do not discuss in detail in this case study. These high-level processing functions tend to be more application specific, and in any case normally represent less than 20% of the overall processing. It should be noted that image acquisition is performed concurrently with processing, and therefore does not affect the timings given in Table 11.3.

11.5.4 Summary

This case study has been concerned with the design of a real-time system for the detection of contaminants in food products. Processing the images in real time constituted a very significant problem, not only in achieving the necessary throughput rate but also regarding system cost and certain other issues.

Careful consideration showed that the most suitable type of real-time solution is software-based, thereby aiding system adaptability and flexibility. Part of the reason for this is that such a system configuration can implement not only necessary convolutions but also, at least in principle, any other relevant algorithms; thus the proposed system would not be restricted, for example, by the capabilities of a convolution board, but would be equally capable of performing morphological operations, or a variety of high-level functions. In general, dedicated hardware solutions embody gross levels of complexity and cost, corresponding to a parallelism which may not be kept 'usefully busy', whereas a software-based system is not subject to this problem. However, the basic processor which is selected for the software solution has to have an instruction set which is both general and which matches sufficiently closely the type of task demanded of it in inspection applications. This is indeed the case with the particular hardware considered in this case study.

11.6 Short case studies of X-ray inspection

11.6.1 Combining X-ray imaging and machine vision (Wagner 1987)

This is one of the earliest papers to show how X-ray imaging can be combined with machine vision for flaw detection in the food industry.

11.6.2 Intelligent visual inspection of food products (Chan et al. 1990)

This paper reviews the problems of machine vision inspection of food products for defects and foreign objects. Use of ultra-violet, infra-red, gamma ray and X-ray inspection modalities is recommended where appropriate. In addition, the value of Prolog for intelligent processing of food images is emphasized, and several short Prolog routines for various food inspection tasks are presented.

A review article by Chan and Batchelor (1993) extends this discussion, and also goes into some detail on the inspection of cakes for cracks and for irregularities and breaks in decorative icing patterns, morphological processing (Chapter 3) being employed for crack detection.

11.6.3 Approaches to foreign body detection in foods (Graves et al. 1998)

This paper reviews the problems of inspecting food products to locate foreign objects. The modalities considered include not only standard vision techniques and their X-ray, microwave, infra-red, and ultrasonic counterparts, but also ultraviolet and fluorescence imaging, nuclear magnetic resonance (NMR) and magnetic resonance imaging (MRI), and the more widely used non-imaging approaches—balanced coil and magnetic metal detectors; the possibility of capacitive imaging is also considered, though the ultimate value of this approach is as yet unclear. Highly relevant is the remark that in the UK, the presence of foreign bodies in food is the largest cause of food defect prosecutions, though (a) many complaints do not get to the stage of prosecutions, and (b) it is not known how many of the complaints are genuine.

11.6.4 Defect detection in apples (Schatzki et al. 1997)

This paper describes preliminary research aimed ultimately at the development of an on-line X-ray inspection station for the detection of defects such as bruises and 'watercore' in apples. Tests were aimed at a commercial target of $\geq 50\%$ defect recognition with $\leq 5\%$ false positives, but the target was only met with some types of apple.

11.6.5 Detection of interior voids, inclusions and pests (Tollner et al. 1993)

This research shows that X-ray detection of voids and inclusions such as glass, metal and stones in objects such as potatoes is possible, though X-ray energies of 40–60 kV are required to penetrate 10–15 cm of food.

11.6.6 Inspection of metal and glass containers (Novini 1990)

This paper addresses the issues involved with inspection of empty cans which are to be used for beverages and foods. Particular attention is paid to completeness of surface coating; presence of scratches, dents, scrap and foreign matter; and (where relevant) presence of pull tag. The work does not involve X-rays but is mentioned here for completeness and for comparison with other papers on packaged foods (see below).

11.6.7 Inspection of packaged foods (Cambier and Pasiak 1986)

This is one of the earliest papers describing the X-ray inspection of packaged foods. Bottle, jar and can packaging are considered, and the flaws detected include metal wires and glass fragments. Typical image processing functions used to accurately locate bottles and other (non-flexible) containers are region–of–interest handling, resampling, intensity remapping to cope with variations in X-ray illumination, spatial filtering, perimeter masking and histogram analysis. With the 43 ms cycle time needed to match the line speed, any more sophisticated detection schemes were apparently ruled out. Illumination from above led to some blind spots just under the shoulder of a jar, but good sensitivity was available for the all-important region at the bottom of a jar. The probability of detection of 3 mm glass beads was found to be 95%, with an accompanying false positive rate of 3%.

11.6.8 Inspection of canned products (Penman et al. 1992)

This paper describes X-ray inspection of canned food products. In this case the (round) cans were viewed from the side, and the difference in absorption over the width of the cans varied correspondingly, so simple thresholding was not a viable option for foreign object detection. In addition, the seams of steel cans sometimes led to problems, causing false alarms. The strategy of processing was to establish a region of interest, to detect pixels of interest, to identify discrete objects, and finally to classify objects as foreign or not. The two basic approaches used were edge detection and morphological processing. In the latter approach, a suitable filter kernel was decided upon for elimination of any dark foreign object pixels, and these were then identified by image subtraction. However, labelling of pixels with suitable parameters permitted better informed decisions about the presence of foreign objects at the classification stage. Tests were made using cans of milk powder and cans of diced apple. Metal wires were located in the milk powder, but the greater absorption of X-rays in the apple resulted in significantly larger pieces of metal being detectable in that case—and correspondingly higher false positive detection rates. In this work 85 kV X-rays were used with 4–6 mA of current to inspect cans of milk powder as large as 2.5 kg (5 litre) and cans of other products (such as apple and beetroot) of up to 3 kg. It is remarked that conventional metal detectors are not well suited to detecting small metal fragments in steel cans, and that X-ray imaging provides the opportunity to solve this problem.

11.6.9 3D inspection of food products (Graves et al. 1994)

This paper summarizes the problems of X-ray inspection of food products to locate foreign objects such as fragments of metal or glass in jars and cans, and pieces of bone in chicken produce, and makes recommendations about the use of several differently orientated X-ray sources and detectors to ensure locating all the foreign objects. This approach is likely to be of value when the area presented to the detector is small, because of the particular orientation of the foreign object, or when it is located under the rim of the container or adjacent to the boundary as seen by one sensor. High pass filtering and use of morphological operators are suggested for highlighting small contaminant bodies.

11.7 Concluding remarks

This chapter has again presented three substantial case studies plus a number of short ones: in this case too, the latter are able to provide a wealth of information on the variety of the subject area, showing that X-ray inspection is of some importance both for ensuring the integrity of the basic product and for confirming, as far as possible, that no foreign bodies exist within it. Nor is this the end of the story, as some X-ray case studies have been included in other chapters—for example, Chapter 10 on cereal grain inspection and Chapter 13 on meat and fish inspection—where they seemed to fit in better. In many cases, as might have been expected, the foreign bodies being detected are fragments of metal and glass—though stones, pieces of rubber, plastic and wood are also sometimes found. In other cases, the foreign bodies are voids or (in the case of potatoes) clods of soil; insects are also detectable and indeed are a serious concern, as they are not static, but will eat away the product during storage.

X-rays are of major importance because they permit the final packaged product to be inspected, at the stage when all the value has been added—and at the same point where any contaminant that could have been included is already present (this ignores the possibility of later deliberate contamination in the market). Bearing in mind that inspection is not cost-free, and assuming that it has to be done, it is better in principle that it is done later than earlier. However, this ignores the not inconsiderable cost of X-ray inspection, which includes the capital cost of the complete inspection installation including all the necessary shielding and safety interlocks. On the other hand, it is becoming necessary for food suppliers to observe 'due diligence', so that if it is technologically possible to check for contaminants and foreign bodies, and can reasonably be expected, then it is incumbent on the supplier to attend to it. Hence it can be assumed that X-ray inspection will gradually assume greater importance and wider use in the coming few years. It is hoped that this chapter will help to make all parties more aware of the situation and of any obligations they have for related public health and safety matters.

Chapter 12
Image processing in agriculture

12.1 Introduction

Agriculture covers an immense range of activities and it will be impossible to do full justice to the subject within the confines of a single chapter. It seemed that the only way in which the right emphasis could be given to the work that has been carried out on the application of image processing in agriculture would be to present a number of case studies which would provide some indication of the possibilities, the underlying principles, and the considerations that must be borne in mind when progressing towards real applications. The case studies that have been selected for this purpose include guidance of crop-spraying vehicles, model-based tracking of animals (for animal husbandry), and inspection and grading of farm produce—specifically potatoes, apples and mushrooms. At the end of the chapter, the various threads are pulled together, and the variety of work on inspection and grading of farm produce that has so far been carried out is summarized.

12.2 Case study: guidance of a crop-spraying vehicle

In recent years, there has been increasing pressure on farmers to reduce the quantities of chemicals used for crop protection. This cry has come both from environmentalists and from the consumers themselves. The solution to this problem lies in more selective spraying of crops. For example, it would be useful to have a machine which would recognize and spray weeds with herbicides, leaving the vegetable crops themselves unharmed: alternatively, the individual plants could be sprayed with pesticides. This case study relates to the design of a vehicle which is capable of tracking plant rows and selecting individual plants for spraying (Marchant and Brivot 1995; Marchant 1996; Brivot and Marchant 1996; Sanchiz *et al.* 1996; Marchant *et al.* 1998).

Figure 12.1 Crops planted in straight lines on a small farm.

The problem would be enormously simplified if plants grew in highly regular placement patterns, so that the machine could tell 'at a glance' from their positions whether they were weeds or plants, and deal with them accordingly. However, the growth of biological systems is somewhat unpredictable and renders such a simplistic approach impracticable. Nevertheless, if plants are grown from seed in a greenhouse, and transplanted to the field when they are approaching 100 mm high, they can be placed in straight parallel rows, which will be approximately retained as they grow to full size. There is then the hope (as in the case shown in Figure 12.1) that the straight rows can be extracted by relatively simple vision algorithms, and the plants themselves located and identified straightforwardly.

At this stage the problems are (1) that the plants will have grown to one side or another, and will thus be out of line; (2) that some will have died; (3) that weeds will have appeared near some of the plants; (4) that some plants will have grown too slowly, and will not be recognized as plants. Thus a robust algorithm will be required to perform the initial search for the plant rows. The Hough transform approach is well adapted to this type of situation: specifically, it is well suited to looking for line structure in images.

The first step in the process is to locate the plants. This can be achieved with reasonable accuracy by thresholding the input images (this process is eased if infra-red wavelengths are used to enhance contrast). However, at this stage the plant images become shapeless blobs or clumps (Figure 12.2). These contain holes and lobes (the leaves, in the case of cabbages or cauliflowers), but a certain amount of tidying up can be achieved either by placing a bounding box around the object shape, or by performing a dilation of the shape which will regularize it and fill in the major concavities (the real-time solution employed the first of these methods). Then the position of the centre of mass of the shape is determined, and it is this that is fed to the Hough transform straight line (plant row) detector. In common with the usual Hough transform approach, votes are accumulated in parameter space for all possible parameter combinations consistent with the input data. Here this means taking all possible line gradients and intercepts for lines passing through a given plant centre and accumulating them in parameter space. To help get the most meaningful solution, it is useful to accumulate values in proportion to the plant area. In addition, it should be noted that if three rows of plants appear in any image, it will not initially be known which plant is in which row, and therefore each plant should be allowed to vote for all the row positions: this will naturally only be possible if the inter-row spacing is known and can be assumed in the analysis. However, if this procedure is followed, the method will be far more resistant to missing plants and to weeds which are initially mistakenly assumed to be plants.

The algorithm is improved by preferentially eliminating weeds from the images before applying the Hough transform. Weed elimination is achieved by three techniques—hysteresis thresholding, dilation and blob size filtering. Dilation refers to the standard shape expansion technique described in Chapter 3 and is used here to fill in the holes in the plant blobs. Filtering by blob area is reasonable since the weeds are seldom as strong as the plants, which were transplanted only when they had become well established.

Hysteresis thresholding is a widely used technique which involves use

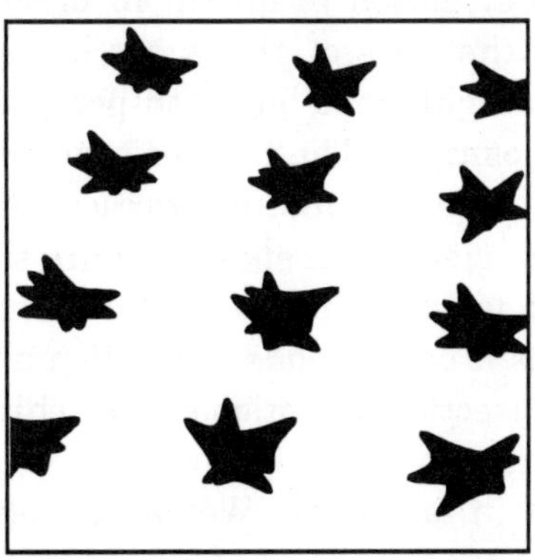

Figure 12.2 Perspective view of plant rows after thresholding. In this idealized sketch, no background clutter is shown.

of two threshold levels. In this case, if the intensity is greater than the upper level t_u, the object is taken to be plant; if lower than the lower level t_l, it is taken to be weed; if at an intermediate level and next to a region classified as plant, it is taken to be plant; the plant region is allowed to extend sequentially as far as necessary, given only that there is a contiguous region of intensity between t_l and t_u connecting a given point to a true plant ($\geq t_u$) region.

Once the Hough transform has been obtained, the parameter space has to be analysed to find the most significant peak position. Normally, there will be no doubt as to the correct peak—even though the method of accumulation permits plants from adjacent rows to contribute to each peak. The reason for this is that with three rows each permitted to contribute to adjacent peaks, the resultant voting patterns in parameter space are: 1,1,1,0,0; 0,1,1,1,0; 0,0,1,1,1—totalling 1,2,3,2,1—thereby making the true centre position the most prominent (actually, the position is more complicated than this as several plants will be visible in each row, thus augmenting the central position further). However, the situation could be erroneous if any plants are missing. It is therefore useful to help the Hough transform arrive at the true central position. This can be achieved by applying a Kalman filter to keep track of the previous central positions, and anticipate where the next one will be—thereby eliminating false solutions. This concept is taken furthest in the paper by Sanchiz *et al.* (1996), where the individual plants are all identified on a reliable map of the crop field and errors from any random motions of the vehicle are systematically allowed for.

12.2.1 3D aspects of the task

So far we have assumed that we are looking at simple 2D images which represent the true 3D situation in detail. In practice this is not so. The reason for this is that the rows of plants are being viewed obliquely and therefore appear as straight lines but with perspective distortions which shift and rotate their positions. The full position can only be worked out if the vehicle motions are kept in mind. In practice, vehicles rolling along the rows of plants exhibit variations in speed and are subject to roll, pitch and yaw. The first two of these motions are less relevant and are ignored here. The last is important as it corresponds to rotation about a vertical axis and affects the immediate direction of motion of the vehicle.

To proceed, we have to relate the position (X,Y,Z) of a plant in 3D with its location (x,y) in an image. We can achieve this with a general translation:

$$T = (t_x, t_y, t_z)^{\mathrm{T}} \tag{12.1}$$

together with a general rotation:

$$R = \begin{bmatrix} r_1 & r_2 & r_3 \\ r_4 & r_5 & r_6 \\ r_7 & r_8 & r_9 \end{bmatrix} \tag{12.2}$$

giving:

$$\begin{bmatrix} X \\ Y \\ Z \end{bmatrix} = \begin{bmatrix} r_1 & r_2 & r_3 \\ r_4 & r_5 & r_6 \\ r_7 & r_8 & r_9 \end{bmatrix} \begin{bmatrix} x \\ y \\ z \end{bmatrix} + \begin{bmatrix} t_x \\ t_y \\ t_z \end{bmatrix} \tag{12.3}$$

The lens projection formulae are also relevant:

$$x = fX/Z \tag{12.4}$$

$$y = fY/Z \tag{12.5}$$

We shall not give a full analysis here, but assuming that roll and pitch are zero, and that the heading angle (direction of motion relative to the rows of plants) is ψ, and that this is small, we obtain a quadratic equation for ψ in terms of t_x. This means that two sets of solutions are in general possible. However, it is soon found that only one solution matches the situation, as the wrong solution is not supported by the other feature point positions. This shows the complications introduced by perspective projection—even when highly restrictive assumptions can be made about the geometrical configuration (in particular, ψ being small).

12.2.2 Real-time implementation

Finally, it was found to be possible to implement the vehicle guidance system on a single T805 transputer operating at 20 MHz coupled with two special hardware units—a colour classifier and a chaincoder. The latter is useful for fast shape analysis following boundary tracking. The overall system was able to process the input images at a rate of 10 Hz, which is sufficient for reliable vehicle guidance. Perhaps more important, the claimed accuracy is in the region of 10 mm and 1° of angle, making the whole guidance system adequate to cope with the particular slightly constrained application considered. A later implementation (Marchant *et al.* 1998) used IMS B429 transputers and did a more thorough job of segmenting the individual plants (though still not using the blob size filter), obtaining a final 5 Hz sampling rate—again fast enough for real-time application in the field. All in all, this case study demonstrates the possibility of highly accurate selective spraying of weeds, thereby very significantly cutting down the amount of herbicide needed for crops such as cabbages, cauliflowers and wheat.

12.3 Case study: model-based tracking of animals

This case study is concerned with animal husbandry—the care of farm animals. Good stockmen notice many aspects of the behaviour of the animals, and learn to respond to them. Fighting, bullying, tail biting, activity, resting behaviour and posture are useful indicators of states of health, potential lameness or heat stress, while group behaviour may indicate the presence of predators or human intruders. In addition, feeding behaviour is all important, as is the incidence of animals giving birth or breaking away from the confinement of the pen. In all these aspects, automatic observation of animals by computer vision systems is potentially useful.

Some animals such as pigs and sheep are lighter than their usual backgrounds of soil and grass, and thus they can in principle be located by thresholding. However, the backgrounds may be cluttered with other objects such as fences, pen walls, drinking troughs, and so on—all of which will complicate interpretation. Whatever the situation on factory conveyors, straightforward thresholding is unlikely to work well in normal outdoor and indoor farm scenes.

McFarlane and Schofield (1995) tackled this problem by image differencing. While differencing between adjacent frames is useful only if specific motion is occurring, far better performance is achievable if the current frame is differenced against a carefully prepared background image. Employing such an image might seem to be a panacea, but this is not so, as the ambient lighting will vary through the day, and care must be taken to ensure that no animals are present when a background image is being taken. McFarlane and Schofield used a background image obtained by finding the median intensity at each pixel for a whole range of images taken over a fair period, in an effort to overcome this problem: intelligent applications of this procedure were instituted to mask out regions where piglets were known to be resting. In such applications, medians are more effective than mean values, as they robustly eliminate outlier values rather than averaging them in.

The McFarlane and Schofield algorithm modelled piglets as simple ellipses, and had fair success with their approach. However, we next examine the more rigorous modelling approach adopted by Marchant and Onyango (1995), and developed further by Onyango and Marchant (1996) and Tillett *et al.* (1997). We group this work together under the heading MOT, to include all three authors.

MOT aimed to track movements of pigs within a pen by viewing them from overhead under not too uniform lighting conditions. The main aim of the work at this early stage was tracking the animals, though, as indicated above, it was intended to lead on to behavioural analysis in later work. To find the animals, some form of template matching is required. Shape matching is an attractive concept, but with live animals such as pigs, the shapes are highly variable: specifically, animals which are standing up or walking around will bend from side to side, and may also bend their necks sideways or up and down as they feed. It is insufficient to use a small number of template masks to match the shapes, as there is an infinity of shapes related by various values of the shape parameters mentioned. These parameters are additional to the obvious ones of position, orientation and size.

Careful trials soon show that matching with all these parameters is insufficient, as the model is quite likely to be shifted laterally by variations in illumination: if one side of a pig is closer to the source of illumination, it will be brighter, and hence the final template used for matching will also shift in that direction. The resulting fit could be so poor that any reasonable 'goodness of fit' criterion may deny the presence of a pig. These factors mean that possible variations in lighting have to be taken fully into account in fitting the animal's intensity profile.

A full-blooded approach involves principal components analysis (PCA). The deviation in position and intensity between the training objects and the model at a series of carefully chosen points is fed to a PCA system: the highest energy eigenvalues indicate the main modes of variation to be expected; then any specific test example is fitted to the model and amplitudes for each of these modes of variation are extracted, together with an overall parameter representing the goodness of fit. Unfortunately, this rigorous approach is highly computation intensive, because of the large number of free parameters; in addition, the position and intensity parameters are disparate measures which require totally different and somewhat arbitrary scale factors to be used to coax the schema into working. This means that some means is required for decoupling the position and intensity information. This is achieved by performing two independent principal components analyses in sequence—first on the position coordinates and then on the intensity values.

When this procedure is carried out, three significant shape parameters are found, the first being lateral bending of the pig's back, accounting for 78% of the variance from the mean; and the second being nodding of the pig's head: as the latter corresponded to only ~20% of the total variance, it was ignored in later analysis. In addition, the grey-level distribution model had three modes of variation amounting to a total of

77% of the intensity variance: the first two modes corresponded to (1) a general amplitude variation in which the distribution is symmetrical about the backbone, and (2) a more complex variation in which the intensity distribution is laterally shifted relative to the backbone (this clearly arises largely from lateral illumination of the animal)—see Figure 12.3.

While principal components analyses yield the important modes of variation in shape and intensity, in any given case the animal's profile has still to be fitted using the requisite number of parameters—one for shape and two for intensity. The Simplex algorithm (Press *et al.* 1992) has been found effective for this purpose. The objective function to be minimized to optimize the fit takes account of (1) the average difference in intensity between the rendered (grey-level) model and the image over the region of the model; and (2) the negative of the local intensity gradient in the image normal to the model boundary averaged along the model boundary (the local intensity gradient will be a maximum right around the animal if this is correctly outlined by the model).

One crucial factor has been skirted over in the preceding discussion: that the positioning and alignment of the model to the animal must be highly accurate (Cootes *et al.* 1992). This applies both for the initial PCA and later when fitting individual animals to the model is in progress. Here we concentrate on the PCA task. It should always be borne in mind when using PCA that the latter is a method of characterizing deviations: this means that the deviations must already be minimized by referring all variations to the mean of the distribution. Thus it is very important when setting up the data to bring all objects to a common position, orientation

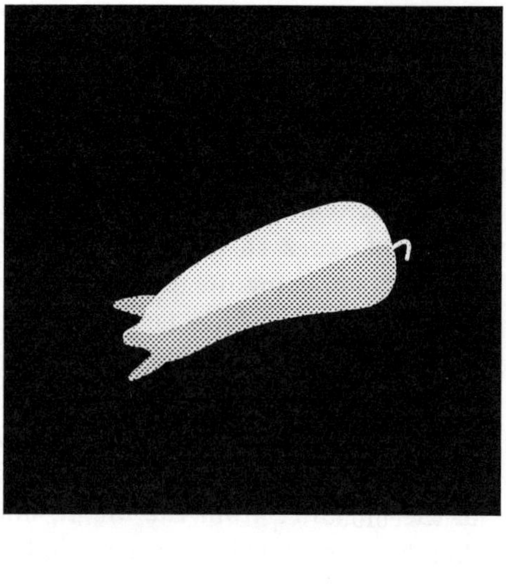

Figure 12.3 Effect of one mode of intensity variation found by PCA. This mode clearly arises from lateral illumination of the pig.

and scale before attempting PCA. In the present context the PCA relates to shape analysis, and it is assumed that prior normalizations of position, orientation and scale have already been carried out. (Note that in more general cases scaling may be included within PCA if required. However, PCA is a computation intensive task, and it is best to encumber it as little as possible with unnecessary parameters.)

Finally, we note that the work of MOT is at an early stage, having been carried out with relatively few images of few animals. Indeed, Tillett *et al.* (1997) admit to taking "seven image sequences of a single pig in a pen", thereby eliminating the need for scale normalization (the pig is viewed by a camera looking vertically down from a fixed height). This should not detract from the achievements of MOT which are worthwhile—particularly in the effective method for decoupling shape and intensity analysis, and show clearly the complexities of the situation. Overall, this work holds out great promise for wide application in animal husbandry, showing that animal monitoring and ultimately behavioural analysis should be attainable over the coming decade with the aid of image processing.

12.4 Case study: inspection and grading of potatoes

Inspection and grading of vegetables is now an important part of the food production scene. Potatoes figure highly in this process as they constitute a considerable fraction of the vegetables eaten in the western world (Figure 12.4). Thus it is not surprising that potato inspection and sorting has been studied by several researchers.

A very important aspect of potato inspection is that of grading by size and shape. Work along these lines was carried out by Marchant *et al.* (1990). It was vital to view the whole of each potato in order to fully assess 3D size and shape. However, instead of using a number of differently orientated cameras to achieve this, a single camera was used, and the potatoes were rotated beneath it by the simple but effective strategy of transporting the potatoes on a conveyor belt containing a series of rotating rollers (Figure 12.5). While the rollers roll forwards, the tops of the potatoes lying on them move forwards but roll backwards. It can be shown that if the potatoes act as perfect non-slipping cylinders, the centre-lines of their upper surfaces will momentarily be stationary—thereby permitting an image of each potato to be built up line by line. This neat arrangement permits the entire surfaces of the potatoes to be examined for blemishes at the same time as their 3D sizes and shapes are assessed. Blemishes are located by quite complex colour image processing using special real-time hardware modules. However, there is a problem with the overall procedure in that potatoes often touch laterally along the lengths of the rollers, making it non-trivial to determine the length of any single potato. While

Figure 12.4 A typical set of market potatoes as they reach the consumer. Notice the defects which are typical of this commodity.

singulation (i.e. separation) of the potatoes would be the ideal, this would demand additional hardware, so a special algorithm was developed to locate the cusps between the potato shapes. This operates on the chain-coded boundary and determines when the distance from a chord to an arc along the boundary is a maximum and is simultaneously greater than a certain threshold value. The threshold and the length of the probing arc are both determined carefully for each size of potato and incorporated into the algorithm. A high degree of reliability was achieved with this machine, and special transputer-based hardware was built to implement it in real time for commercial practice (Marchant and Tillett 1994).

In a separate set of studies, Heinemann *et al.* (1996), Tao *et al.* (1995a) and Tao *et al.* (1995b) developed separate methods for potato inspection and grading. These papers will be considered together in the remainder of this case study. First, a study was made of methods for

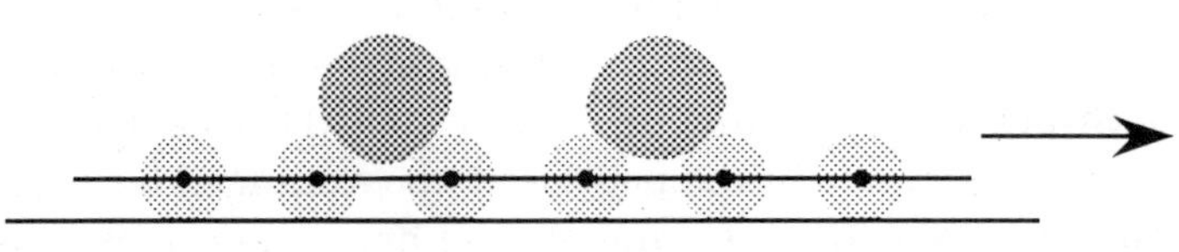

Figure 12.5 Potatoes being transported on a conveyor containing a series of rolling rollers. Ideally, the top line of each potato is momentarily stationary.

analysing the shapes of potatoes. Curvature, compactness, bending energy, moments and several other measures were investigated but were found wanting in one way or another: in particular, it was found that they were unable to separate shapes as required, because different shapes sometimes gave coefficients with the same values. However, the Fourier boundary descriptor technique was found suitable if the first ten harmonics were used: these ten harmonics were able to reproduce actual potato shapes sufficiently accurately (reproduction being a criterion *par excellence* for judging adequacy of shape representation). Nevertheless, cluster analysis of these ten coefficients was not found effective for potato grading—presumably because of the lack of statistical significance of so many coefficients. Accordingly, a simple shape heuristic separator which weighted the effects of the higher harmonics appropriately was devised and used as the basis for the grading system. The final system was able to discriminate successfully between 'fairly well shaped', 'not seriously misshapen', and 'seriously misshapen' samples, using two thresholds of 1.0 and 1.75 for the single Fourier shape heuristic. The overall system was able to process potatoes at the rate of three per minute, which is prohibitively slow, though it could in some cases be adequate for grading based on sampling.

12.4.1 Colour inspection of potatoes

Colour inspection of potatoes follows a similar line to that for apples (see Section 12.5.2). The major task of potato colour inspection is to check for surface greening, which is largely due to sunburn. Colour assessment is carried out via the hue histogram. However, a simple thresholding approach would not achieve the desired results, as there is a considerable degree of overlap between the histograms of greened and non-greened potatoes. A neat solution to this problem was found by splitting the central active range of the hue histogram, normalized to fixed object size, into two sub-ranges, and using the numbers of pixels in each of these ranges as features of a 2D discriminant space. The clusters in this space were found to be separable between the two classes, and 2D linear discriminant analysis was used for classification. This approach resulted in misclassification rates of 5–10%. Some complications arose from potatoes with black or white spots, those with white-yellow blemishes, and those with mechanical damage, but these do not invalidate the basic methodology. Indeed, the 2D discriminant analysis procedures devised in this work seems general and should be applicable to other kinds of fruit and vegetables.

12.5 Case study: inspection of apples

Apples represent one of the most important fruit tree crops in the western world and present a formidable inspection problem. Human inspectors need to be trained if they are to meet to meet the required standards (e.g. those laid down by USDA in the USA), and inevitably they become tired, inconsistent and unreliable during long shifts of working. In any case, human estimates of quantities such as defect area can be very subjective. These factors have impelled workers to investigate the design of machine vision inspection systems which are both effective and completely consistent in their examination of apples and other fruit.

Particular targets for inspection are shape, colour and defects; the latter include russetting, bruising, unhealed cuts, decay, worm holes, insect injury and injury induced by weather (especially hail). Russet is a type of cork that grows in a net-like pattern on the surface of apples and is one of the most prevalent defects of the Golden Delicious variety, bruising being another. Throop and Aneshanesley (1993) have studied bruise detection. This case study describes the work of Heinemann *et al.* (1995) on apple inspection, and includes the topics of shape analysis, colour and russetting. It then goes on to consider the approach taken by Yang and Marchant (1996) to assess defects or 'blemishes' using snake algorithms.

12.5.1 Shape

One of the most important shape defects of apples is lop-sidedness (Figure 12.6). This may be quantified by the maximum/minimum length ratio, the relevant maximum and minimum heights being measured (ignoring the stem) by resting the apple on its calyx (Kuhn *et al.* 1982). It was found that presence of the stem prevented simple boundary analysis from working reliably, so the whole cross-sectional area was analysed by moments[1]. As lop-sidedness is akin to skewness, the latter is only observable via the odd order moments; thus the third order moment about the vertical axis was used for this purpose. In a sample of 20 well-shaped and 20 misshapen applies, this approach led to 92% agreement between human and machine grading in the detection of misshapen apples, and a threshold change of ± 5% led to only 1.1% additional misclassifications. Five views were recommended to guarantee achieving the required level of accuracy.

---

[1] It seems likely that it would be more computationally efficient to analyse shape by boundary rather than cross-sectional area analysis if only the stem could be eliminated from consideration. This might reasonably be achieved by outlier elimination in a running 1D window used to filter the boundary information.

Figure 12.6 Geometry for characterizing lop-sidedness of apples.

12.5.2 Colour

Colour plays a big part in the determination of apple quality. Not only is it a prime indicator of ripeness, but also it contributes greatly to physical attractiveness, and thus induces purchase and consumption. In addition, the USDA (and almost certainly other national bodies) has very specific requirements for the colour range of apples: thus US fancy Golden Delicious should have 75% or more of the surface showing white or light green predominating over green, while only a quarter of apples are permitted to show significant 'blush'.

Whereas colour cameras digitize colour into the usual RGB (red, green, blue) channels, humans perceive colour differently. As a result, it is better to convert the RGB representation to the HSI (hue, saturation, intensity) domain before assessing the colours of apples and other products[2]. Space prevents a detailed study of the question of colour: the reader is referred to more specialized texts for detailed information (e.g. Gonzalez and Woods 1992; Sangwine and Horne 1998). However, some brief comments will be useful. Intensity I refers to the total light intensity, and is defined by:

$$I = (R + G + B)/3 \tag{12.6}$$

Hue H is a measure of the underlying colour, and saturation S is a measure of the degree to which is it diluted by white light. S is given by the simple formula:

[2] Usually, a more important reason for use of HSI is to employ the hue parameter which is independent of the intensity parameter, as the latter is bound to be particularly sensitive to lighting variations.

$$S = 1 - \frac{\min(R,G,B)}{I} = 1 - \frac{3\min(R,G,B)}{R+G+B} \tag{12.7}$$

which makes it unity along the sides of the colour triangle, and zero for white light ($R = G = B = I$).

Hue is defined as an angle of rotation about the central white point W in the colour triangle: it is the angle between the pure red direction and the direction of the colour in question. The derivation of a formula for H is quite complex and will not be attempted here. The final result is:

$$H = \cos^{-1}\left(\frac{\frac{1}{2}[(R-G)+(R-B)]}{\left[(R-G)^2 + (R-B)(G-B)\right]^{1/2}} \right) \tag{12.8}$$

or 2π minus this value if $B > G$ (Gonzalez and Woods 1992).

When checking the colour of apples, the hue is the important parameter. A rigorous check on the colour can be achieved by constructing the hue histogram distribution and comparing it with that for a suitable training set. The most straightforward way to carry out the comparison is to compute the mean and standard deviation of the two distributions to be compared and to perform discriminant analysis assuming a Gaussian distribution function. This leads to the formula:

$$p_i(x) = \frac{1}{(2\pi\sigma_i^2)^{1/2}} \exp\left[-\frac{(x-\mu_i)^2}{2\sigma_i^2} \right] \tag{12.9}$$

where the suffix i refers to the two respective distributions. However, applying the respective *a priori* class probabilities P_1, P_2 (Chapter 7), careful analysis (Gonzalez and Woods 1992) shows that the condition $p_1(x) = p_2(x)$ reduces to the form:

$$x^2\left(\frac{1}{\sigma_1^2} - \frac{1}{\sigma_2^2}\right) - 2x\left(\frac{\mu_1}{\sigma_1^2} - \frac{\mu_2}{\sigma_2^2}\right) + \left(\frac{\mu_1^2}{\sigma_1^2} - \frac{\mu_2^2}{\sigma_2^2}\right) + 2\log\left(\frac{P_2\sigma_1}{P_1\sigma_2}\right) = 0 \tag{12.10}$$

Note that, in general, this equation has two solutions, implying the need for two thresholds, though when $\sigma_1 = \sigma_2$ there is a single solution:

$$x = \frac{1}{2}(\mu_1 + \mu_2) + \frac{\sigma^2}{\mu_1 - \mu_2}\ln\left(\frac{P_2}{P_1}\right) \tag{12.11}$$

In addition, when the prior probabilities for the two classes are equal, the equation reduces to the altogether simpler and more obvious form:

$$x = \frac{1}{2}(\mu_1 + \mu_2) \tag{12.12}$$

Returning to the work of Heinemann *et al.* (1995), discriminant analysis of colour based on the above theory gave complete agreement between human inspectors and the machine following training on 80 samples and testing on another 66 samples. However, a warning about maintaining the lighting intensity levels identical to those used for training was given: in any such pattern recognition system it is crucial that the training set be representative in every way of the eventual test set.

12.5.3 Russetting

The degree of russetting could be identified and measured by the same procedure as for colour. Again a linear discriminant function was trained, this time on 32 samples, and tested on another 40 samples. This time 82.5% accuracy was attained relative to human grading. The accuracy of russet grading was lower than that for normal colour grading partly because human inspectors had to estimate the area of russet before assigning a grade, and human estimates of area can be very subjective. It therefore seems that, in this case, the absolute error rate of the computer is closer to 2.5%.

12.5.4 Blemish detection

While certain types of defect will be detectable only by colour, there are many types of blemish that are readily detectable using normal grey-scale imaging. This applies especially to light yellow or green apples such as Golden Delicious. In such cases thresholding is an obvious approach for segmentation of dark defects including worm holes, scabs, bruises and prominent russet patches. However, in most instances it is impossible to set a single threshold value for segmentation. A good part of the problem is the variation in lighting over the object, whose local orientation varies widely over its visible surface, and whose absolute reflectance also varies significantly. In early researches, Yang (1994) developed a 'water catchment basin' model of the blemishes, in which water is imagined accumulating in water catchment basins (the darkest regions) and gradually flooding into wider and wider basins (slightly lighter regions). When the widest such regions (excluding the whole of the apple) have been reached, they are taken to represent the blemishes. This approach proved more effective than earlier approaches such as the background subtraction method of Rehkugler and Throop (1989) and the thresholding method (Davenel *et al.* 1988). However, the segmentation it produced was somewhat coarse in boundary localization: this prompted use of median and Gaussian filters to refine the results. In fact, median filtering did help marginally by removing impulse noise, but Gaussian smoothing had the effect of blurring the boundary so that when a threshold was eventually applied, the size of the blemish was often increased in size: this stems from the fact that a Gaussian mixes signal (the blemish) into further off regions where it is absent, thereby increasing the apparent size.

To overcome these problems, the 'snake' (active contour) type of algorithm was tried. The idea of a snake is to slither around finding a position which in some sense minimizes boundary energy (Kass and Witkin 1987; Kass *et al.* 1988). Usually, snakes are closed loops rather than open-ended arcs, so the concept identifies well with optimization of a blemish boundary (Figure 12.7). A useful further analogy is with an elastic band which fits around the blemish, though this may convey the idea that the final blemish boundary will shrink too far—an unnecessary connotation, though one that is possible if the mathematical model used includes appropriate parameter values.

The energy of the snake is represented by the sum of three main components—the snake's internal energy and the external energy supplied by the image data and by the constraints imposed by the model. Perhaps the most important type of constraint is that preventing the snake from moving into prohibited regions, such as beyond the image boundary, or, for a moving vehicle, into building regions—or, in the present case, outside the confines of an apple. Alternatively, attraction would be possible via a sign change in the relevant weighting factor.

The snake's internal energy includes elastic energy which might be needed to extend or compress it, and bending energy. If no bending energy

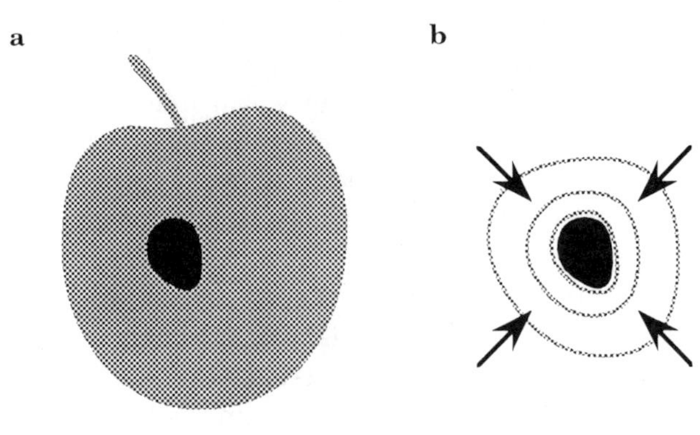

Figure 12.7 Use of snakes to locate blemishes on apples. (a) Apple with blemish. (b) How the snake gradually zones in on the single dark blemish.

terms were included, sharp corners and spikes in the snake would be free to occur with no restriction. Similarly, if no elastic energy terms were included, the snake would be permitted to grow or shrink without penalty.

The image data is normally taken to interact with the snake via three main types of image feature—lines, edges and terminations (the latter can be line terminations or corners). Various weights can be given to these features according to the behaviour required of the snake. For example, it might be required to hug edges and go around corners, and only to follow lines in the absence of edges: so the line weights would be made much lower than the edge and corner weights.

These considerations lead to the following breakdown of the snake energy:

$$\begin{aligned} E_{\text{snake}} &= E_{\text{internal}} + E_{\text{external}} \\ &= E_{\text{internal}} + E_{\text{image}} + E_{\text{constraints}} \\ &= E_{\text{stretch}} + E_{\text{bend}} + E_{\text{line}} + E_{\text{edge}} + E_{\text{term}} + E_{\text{repel}} \end{aligned} \tag{12.13}$$

The energies are written down in terms of small changes in position $\mathbf{x}(s) = (x(s), y(s))$ of each point on the snake, the parameter s being the arc length distance along the snake boundary. Thus we have:

$$E_{\text{stretch}} = \alpha(s)\left|\mathbf{x}_s(s)\right|^2 \tag{12.14}$$

and

$$E_{\text{bend}} = \beta(s)\left|\mathbf{x}_{ss}(s)\right|^2 \tag{12.15}$$

where the suffices s, ss imply first and second order differentiation respectively. Similarly, E_{edge} is calculated in terms of the intensity gradient magnitude $|\text{grad}\, I|$, leading to:

$$E_{\text{edge}} = -w_{\text{edge}} \int |\text{grad}\, I|^2\, ds \tag{12.16}$$

where w_{edge} is the edge weighting factor.

The overall snake energy is obtained by summing the energies for all positions on the snake: a set of simultaneous differential equations is then set up to minimize the total energy. Space prevents us from discussing the full details of this process here. Suffice it to say that the equations cannot be solved analytically, and recourse has to be made to iterative numerical solution, during which the shape of the snake evolves from some high energy initialization state to the final low energy equilibrium state, defining the contour of interest in the image.

12.5.4.1 Problems with snakes

In the general case, there are several possible complications to be tackled: (1) several snakes may be required to locate an initially unknown number of relevant image contours; (2) the different snakes will need different initialization conditions; (3) snakes will occasionally have to split up as they approach contours which turn out to be fragmented.

There are also procedural problems. The intrinsic snake concept is that of well-behaved differentiability. However, lines, edges and terminations are usually highly localized, so there is no means by which a snake even a few pixels away could be expected to learn about them and hence to move towards them. In these circumstances the snake would 'thrash around', grabbing what information it could in a sequential *ad hoc* manner which would not permit it to attain a global minimum of energy. To overcome this problem, gross smoothing of the image is called for, so that edges can communicate with the snake some distance away; and the smoothing must gradually be reduced as the snake nears its target position.

These latter procedures are evident in the Yang and Marchant (1996) work, but no measures appear to have been taken to cope with the previous three complications—probably because they rarely if ever arose in the test data. Nevertheless, it is easily imaginable that two or three quite tiny defects could appear close to each other on the surface of the apple, and the snake would slither into a contour around them all and be unable to fragment itself in such a way as to give a correct interpretation of the situation: interestingly, the presence of a stretching energy term would tend to lead to a triangular solution (Figure 12.8) rather than one in which the enclosed area is minimized (the latter could be the one representing the situation most meaningfully). On the other hand, post-processing to analyse the meaningfulness of the snake data could perhaps lead to solution of such problems.

Finally, the use of the elastic band analogy implies that the snake has a natural size which makes it resist being made larger. We could also imagine it resisting being made smaller. However, it is by no means obvious that apples would have a standard size of blemish as implied by this model. Overall, it appears that only very loose size constraints should be used for detecting blemishes: specifically, a blemish would appear to mean something that is between 2% and 70% of the diameter of the apple, rather than something between 10% and 20% of its size. Nevertheless, the Yang and Marchant (1996) states that the size of the tensile coefficient had little effect on the final snake contours.

Overall, the solutions found by the snake algorithm used in this case study seemed excellent—perhaps partly because of the sound basis provided by the earlier flooding algorithm to initialize the snake algorithm. The caveats given above would be necessary if suitable initializations were

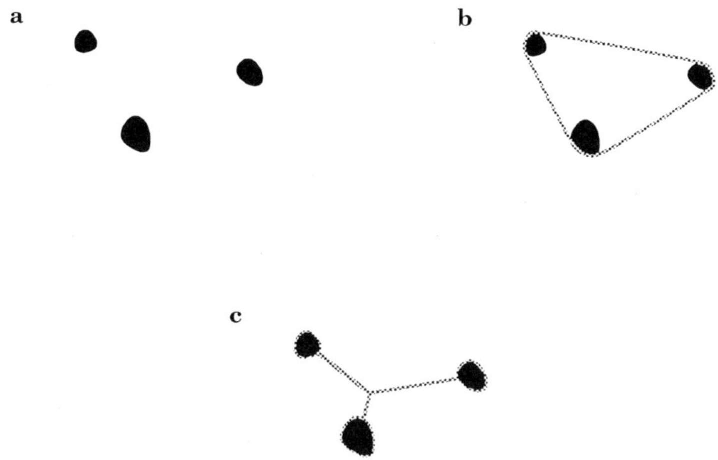

Figure 12.8 Result of applying a snake when three small blemishes appear close together. (a) The three blemishes. (b) The result of applying a snake when the snake energy depends on its length. (c) The likely result of applying a snake when the snake energy depends on the area it encloses. These results highlight the potential problems of using snakes, case (b) being arguably more successful than (a).

not available, or in totally different types of situation such as locating animals in outdoor scenes where background clutter would provide a good many distracters which would almost certainly prevent the snake procedure from working: at the level where the algorithm has no appreciation of the overall situation, but merely reacts to a conglomerate of local pieces of information in the image, successful segmentation using snakes is likely to prove risky despite the intuitive attractiveness of the concept.

12.6 Case study: inspection and grading of mushrooms

This case study is concerned with the inspection and grading of mushrooms. It outlines the work of Heinemann *et al.* (1994), and contains few surprises, as it follows similar lines to those adopted by the authors in their other work on potatoes and apples (see above). One of the most important factors is that standards for grading *Agaricus bisporus* mushrooms were laid down long ago by the USDA (1966)—even though Heinemann *et al.* report that the two-tiered grading standards are rarely, if ever, used in the commercial marketplace. However, commercial mushroom growers in the USA may well soon establish a new more realistic grading standard that will include more grades.

The inspection system of Heinemann *et al.* was designed to detect four categories of defect: coloration, shape, stem cut, and cap veil opening. In this case 'coloration' meant any deviation from white, and thus grey-scale imaging was adequate for the purpose. Specifically, pits and discoloured areas had to be located. Diffuse but uniform lighting was required for this purpose, and this was achieved by use of a white dome within which light tubes were placed: direct light was prevented from reaching the sample mushrooms by use of diffusers, and by diffuse reflection from the dome. Location of individual pits or discoloured regions was not carried out: instead, the shapes of the intensity histograms were analysed. It was found that there was a sharp edge to the left of the histograms, demarcating the intensity of the darkest parts of the mushroom, and that this moved to the left as the mushrooms aged and became darker over just three days. The effect could be quantified by histogram mean and skewness measures, and good separation of acceptable and poor quality mushrooms was achieved, the misclassification rate being around 20%.

The 'shape' of the mushroom refers to the roundness of the cap. This was measured by the method of moments, based on the definitions:

$$m_{pq} = \sum_i \sum_j i^p j^q I(i,j) \qquad p,q = 0,1,2,\ldots \tag{12.17}$$

Invariant moments have been devised which are invariant to translation, rotation and scale change. Of these:

$$\varphi_1 = \mu_{20} + \mu_{02} \tag{12.18}$$

$$\varphi_2 = (\mu_{20} - \mu_{02})^2 + 4\mu_{11}^2 \tag{12.19}$$

were used to quantify the deviation from circularity. In fact, it was found that use of either one of these on its own gave an adequate measure of circularity for the purpose of grading mushrooms, the misclassification rate again being around 20%.

'Veil opening' refers to the opening up of the underside of the cap thereby exposing the gills of the mushroom. This feature shows that the mushroom is starting to degrade and is not present in fresh mushrooms of this variety. It is measured by the area of the mushroom, viewed from below, which is below a brightness threshold. The misclassification rate on this test was around 16%.

Finally, the 'stem cut' is analysed, starting from a side view of the mushroom. This leads to three measures: the length of the stem, the flare of the stem, and the angle of cut at the bottom of the stem. Note that these three measures indicate the skill of the harvester rather than any factor in the growing process. Although flare is mentioned in the USDA grading standards, it seemed that human inspectors ignore it and concentrate on the other two measures—stem length and stem angle. Classification therefore concentrated on a 2D thresholding analysis of these two parameters. The misclassification rate on this combined test was around 34%.

A significant problem in the investigation was the level of disagreement between human inspectors. This arose because of the lack of a realistic consistently recognized standard for mushroom classification. This meant that when the system was trained by one inspector and tested by another, significantly larger misclassification rates emerged. The median misclassification rate for the system was 20%, with discrepancies of up to an additional 36% being ascribable to the inconsistencies of individual human inspectors. A salutary lesson thus emerges from this case study: quality can only be maintained relative to pre-ordained standards, and if these are lacking, machine vision inspection has to aim at a moving target, with the consequence that it is very liable to prove uneconomic. Contrariwise, it is when judgements are made absolute and are rigorously adhered to that machine vision inspection comes into its own and eliminates subjective human judgemental factors and the inconsistencies that arise from boredom, distraction and tiredness.

12.7 Concluding remarks

This chapter has presented a number of case studies representing a range of investigations and examples of image processing applied to agriculture and agricultural produce. The case studies have been chosen to be varied, though it is also clear that there is a class of inspection and grading applications which have significant similarities to each other—while also being subject to the parochial tendencies of the various sectors of the food industry to which they belong. Thus automated mushroom inspection has been limited by the lack of realistic grading standards, while apple inspection has been dominated by the overwhelming importance of physical appearance to the consumer, who will only pick up and eat an apple if it possesses an attractive coloration and is defect free.

There are a good number of products that could be inspected and graded automatically that have not been—or rather, which have not yet been mentioned specifically in the scientific literature. Of those that have, it has not proved possible to cover all of them in the space of a single chapter. Arguably, it is better to cover a few examples in depth, than to race past every possible item rather shallowly: at least a more in-depth approach airs a reasonable number of the major principles and provides a useful backcloth to the discussion. That is not to say that some major principles have not been aired, though it is probably also true to say that the food industry is still progressing in a rather unambitious way, partly because profit margins are generally fairly modest; and case studies for relevant products can hardly be presented before image processing techniques have been applied to them. In these circumstances the best that can be done is to make food production managers aware of the possibilities for applying image processing in the food industry, and this is attended to both via the case studies presented already and by the run-through of major image processing methods covered in Part 1.

Next, it is worth pointing to the literature of the subject and outlining what has been achieved by earlier workers. Many papers have been written that relate to fruit inspection and grading—in particular:

- apples: Graf *et al.* 1981; Kuhn *et al.* 1982; Taylor and Rehkugler 1985; Rehkugler and Throop 1985, 1989; Davenel *et al.* 1988; Varghese *et al.* 1991; Yang 1992, 1994; Throop and Aneshanesley 1993; Heinemann *et al.* 1995; Tao *et al.* 1995a; Yang and Marchant 1996
- bananas: Liao *et al.* 1990
- citrus fruits: Johnson 1985; Maeda 1987; Chen *et al.* 1992
- dates: Wulfsohn *et al.* 1989
- kiwifruit: Hodgson and McNeill (1986)[3]; Bailey and Hodgson (1988)
- oranges: Dennis *et al.* 1988
- peaches: Miller and Delwiche 1989a,b
- prunes: Tang *et al.* 1989; Delwiche *et al.* 1990a,b
- tomatoes: Sarkar and Wolfe 1985a,b, 1986.

Similarly, a great deal of work has been carried out on vegetable (and mushroom) grading:

- asparagus: Rigney *et al.* 1992 (here the detection of defects such as spreading tips, broken tips, stalk scars and cracks is of particular interest)
- bell peppers: Wolfe and Swaminathan 1987; Shearer and Payne 1990
- carrots: Howarth and Searcy 1992 (here the detection of forking, surface defects, breaks and curvature is of particular interest)
- mushrooms: Tillett and Batchelor 1991; Heinemann *et al.* 1994
- potatoes: McClure and Morrow 1987; Marchant *et al.* 1988, 1990; Tao *et al.* 1990; Marchant and Tillett 1994; Tao *et al.* 1995a,b; Heinemann *et al.* 1996;
- processed vegetables: D'Agostino 1990

---

[3] This early work deserves careful study, but details could not be included here for reasons of space: the methods used broadly parallel those for apple inspection, and the 'dropped shoulder' defect is similar to the lop-sidedness defect for apples. Note, however, that the kiwifruit is elongated, unlike the apple which is approximately spherical.

- soybeans: Wigger *et al.* 1988
- sweet potatoes: Wright *et al.* 1984.

It is also relevant to point to some early publications on standards for grading:

- MAFF 1973: EEC standards for fresh apples and pears
- USDA 1966: US standards for grades of mushrooms
- USDA 1978: Apples—market inspection instructions
- USDA 1991: US standards for grades of potatoes.

To a large extent, analysis of size, shape and colour is a standard component of much of the work: a lot is now known about these aspects of inspection and grading, both in general and in specific instances of fruit and vegetables as cited in the different papers. The major problem is often the defects and blemishes, which can take a variety of different forms, and which also appear in a range of sizes, shapes and colorations (many, however, being dark in nature). There is also the problem of distinguishing dark local defects from stalks and calyxes (Kuhn *et al.* 1982; Yang 1994)—which are always present and which they often resemble closely: this is especially relevant when non-orientated apples or pears are being graded.

In addition, selective crop spraying by land-based vehicle is still in its infancy, though significant progress has been made, while animal husbandry is an obvious growth area, with potential for behavioural (even psycho-) analysis of animals to ensure well-being and good growth rates. These latter two topics are the subject of two substantial case studies early in this chapter, and take the subject out of the realms of constrained mass-inspection into a more cerebral area. Nevertheless, the mass-inspection side needs to be taken further, via automatic training methods which adapt straightforwardly from one commodity to another and from one variety to another. In earlier times the lack of cheap hardware systems capable of operating in real time to inspect 100% of products at the necessary rates of 20–100 items per second (for fruit such as apples and vegetables such as potatoes) was a definite problem, and algorithms had to be cut to the bone in order to achieve real-time performance. It can be envisaged that this problem will be significantly reduced over the next 10 years, and by then the lack of inspection standards will be the factor which acts to limit the application of rigorous grading of agricultural produce. Clearly, it is one function of this book, and of this chapter in particular, to raise this problem, and to encourage the food industry to devote some energy to solving it. With due pressure from government and from consumers, progress will occur on an increasing scale for some time to come.

Finally, there is a further area of application of image processing in agriculture which could not be covered in depth in this chapter for reasons of space, though it is likely to become increasingly important in the coming years: this is vision-guided cutting, pruning, planting and micropropagation. In this type of work, parts of plants such as root clumps, branches, and junctions have to be recognized and cutting or planting actions initiated. Some help can be obtained by perceived convexity (including intensity variations), boundary node analysis, medial line estimation, skeleton junction location, and appropriate use of artificial neural networks for judging optimal cutting planes. For work of this type see Tillett (1990); Tillett *et al.* (1990); Tillett and McFarlane (1991); Davis and Tillett (1994).

Chapter 13

Vision for fish and meat processing

13.1 Introduction

This chapter can be regarded as an extension of the previous chapter to cover fish and meat processing issues. A case study approach is again adopted as the most equitable way of covering a large range of possible handling operations and to make best use of the space available. In this case two slightly dated books (Pau and Olafsson 1991; Khodabandehloo 1993a) cover aspects of the subject: hence it will be inappropriate to present too many case studies on overlapping topics. Those that are presented in this chapter are deemed to be highly relevant to the state of the art, and indeed contribute significantly to the scientific base of the subject. The case studies that have been selected include species sorting of fish, grading of prawns and inspection and grading of poultry parts. At the end of the chapter, the variety of work on inspection and grading in this area is summarized, and the main principles and problems are highlighted.

13.2 Case study: species sorting of fish

Commercial fishing is a laborious, labour-intensive job which takes place in all weathers and involves not only handling nets containing fish but also all the accompanying sorting and processing operations. Sorting is necessary because of permitted fish quotas and the need to return non-permitted catch and undesired species quickly to the water—not least in order to overcome the problem of over-fishing. In Europe, the whole process is subject to EU regulations (EC Commission Regulation 3703/85 1986).

Up to now, the sorting process has been carried out by hand, but there is a need for reliability and for real-time handling, which in this case means sorting fish at rates of around 60 fish/minute, or 1 per second. This case study describes research aimed to achieve reliable processing at these rates (Strachan *et al.* 1990; Strachan 1993, 1994): currently the system has led to classification accuracies of 99% and rates of 40 fish/minute: with the increase in computational power currently occurring (roughly, a doubling of computer power every 18 months), this particular problem appears to be solved. Nevertheless, there are many enhancements that might be made to such a system to make it more adaptable and cost-effective, and the principles should be relevant to a good many other food sorting applications.

The basis of the approach is to sort the fish by shape and colour. However, it is first necessary to locate the fish and differentiate them from debris that may be lying on the conveyor. This is carried out by simple probing and checking along a few sampling lines within the image, perpendicular to the direction of motion of the conveyor. Next, the shapes of the fish are analysed by simple shape descriptors which crudely sort the fish. Aspect ratio (length/width, in a position where width is a maximum) and 'area ratio' (area of front half/area of back half of fish, measured along the length) were used for this purpose. Such measures are useful for quickly checking orientation. However, they are rudimentary measures, and a much more rigorous approach to shape analysis is required. The situation is made considerably more complex by the fact that the fishes' body may be bent.

The approach taken is first to find an approximation to the 'skeleton' (Chapter 3). This is achieved by bisecting the fish shape vertically at ten equally spaced intervals along the length, excluding the tail (that part beyond the narrowest point of the fish—see Figure 13.1(a)). The skeleton is used as a basis for subsequent analysis, as it is ideally a curved line along the centre of the body of the fish, and is almost invariant to the distortions caused by bending: thus it makes a good reference from which to initiate other measurements.

Next, normals are drawn from the skeleton joins to the boundaries of the fish (Figure 13.1(b)). In addition, each normal is bisected and the fish is thus divided laterally into four strips of locally equal widths. The result is a shape grid for the fish shape which takes full account of the fishes' orientation and bending. This approach is satisfactory for round fish such as whiting, but is unsuited to analysis of the shapes of flat fish such as lemon sole. The reason for this is that the skeleton normals sometimes cross over, leading to a confusing set of dividing lines (Figure 13.1(c)). However, flat fish bend very little, so it is reasonable in that case to divide their shapes into regions using vertical lines rather than skeleton normals (Figure 13.1(d)). The aspect and area ratios are especially useful as they permit the fish to be categorized as flat or round, as well as helping with head and tail identification and preliminary orientation.

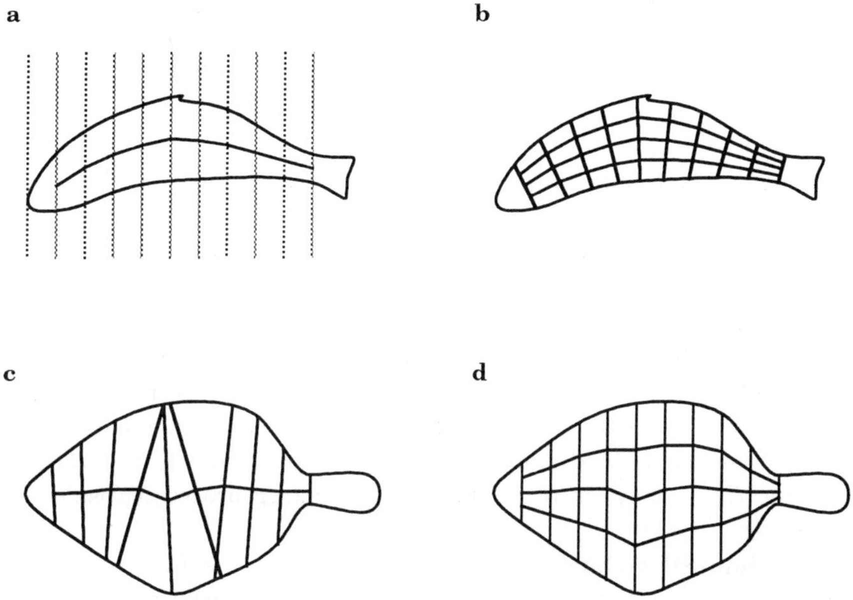

Figure 13.1 Characterization of fish shapes. (a) shows how an approximate skeleton is formed on a fish shape. (b) shows how the skeleton normals may be used to generate 36 central regions, plus the head and tail. (c) shows the result of attempting the same approach for a flat fish: (d) shows an alternative approach to the problem.

Curiously, the only information from the shape grid that is used to characterize the shape of the fish is the set of ten widths of the fish and the length—all these being normalized by dividing by the square root of the fishes' area. However, the shape grid is valuable for providing a detailed colour map of the fish, by averaging the R, G, B values over each of the 36 central grid regions, plus the head and tail.

Discriminant analysis (Chapter 7) was used to analyse the shape and colour data. In fact, it proved necessary to make the multivariate analysis more reliable by reducing the number of parameters. This was achieved by grouping the shape grid regions into three sets—for the front, middle and tail third of the fish—and averaging the R, G, B values over these three regions. This left eleven shape descriptors and three sets of R, G, B values, with the eleven shape descriptors being used together for testing. The four sets of data gave classification accuracies of 94–95% for flat fish, but when used in a majority voting scheme, 100% classification accuracy was achieved. Similar results were obtained for round fish, ending again with 100% overall classification accuracy. With pelagic fish (herring, sprat, mackerel, blue whiting and scad), the results were not dissimilar, but the overall species sorting accuracy was just 98%. The reason for this was that sprat were not always classified correctly, having even probability of being classed as sprat or herring. Nevertheless, this is a notable degree of success. An obvious question is whether human inspectors would ever be confused by sprat, but no information was given about this.

13.3 Case study: grading of prawns

In Australia, prawns are an important commodity, and are marketed in packs of 5–10 with all prawns being roughly the same size and pointing in the same direction. They are caught by trawlers in coastal waters, and sorted and frozen on board. At the on-shore facilities, the batches of prawns are unloaded, thawed, sorted, graded, repacked and refrozen. Multiple freezing and unfreezing operations lower quality, and ideally prawns should be sorted, packed and frozen once on board the trawler. This case study describes research to automate the grading process (Kassler 1990; Kassler *et al.* 1991).

Mechanical technology is able to separate the prawns ready for machine vision analysis: the occasional pairs of entangled prawns can be dealt with by human operators, which would not be entirely dispensed with. The machine vision system has to perform the following tasks:

- locate the prawn on the conveyor
- grade the prawn according to estimated mass
- find the prawn's head, legs, back and tail
- estimate the curl of the prawn
- determine where a robot gripper should pick it up.

Amusingly, the robot system is required to straighten a curled prawn, with the aid of three grippers—a process that takes about 2.5 s. While the machine vision analysis is very fast (around 0.05 s), it is not matched by the robot handling rate. Nevertheless, a single robot working on a two-shift basis should be able to carry out the arduous sorting and handling task previously carried out by 2–3 people working a single shift.

The image processing used to analyse prawn images involves standard operations such as thresholding to binary, connected components analysis, centroid computation, and straightforward shape analysis of the type described in Chapter 3. Given that the prawns are guaranteed to be separated, and will easily threshold to single binary objects, and the fact that size is the single grading parameter, the novelty of this work is predominantly on the robotics side. This illustrates that meat handling is not itself problematic: the difficulties arise (a) when meat objects have complex 3D shapes leading to highly non-uniform lighting (see below), and (b) their surfaces are highly patterned—especially by variegated flesh, fat, skin and bone tones.

Very similar work has been carried out by Krueger (1991), in which the three robot 'grippers' are again suction pads.

13.4 The problem of meat processing

In general, meat offers a serious challenge to the image processing system designer because of the complex and often large shapes that are involved. In this it is in marked contrast with the requirements for inspection of vegetables, fruit, and other such commodities which tend to be small, compact, convex, and of uniform brightness or colour. In particular, when objects have complex shapes, they tend to be self-occluding, or else to cause shadows to fall over their other parts: this means that operations such as thresholding are severely restricted in their capabilities.

For example, large carcasses of beef which have been eviscerated offer complex shapes and colour patterns, and while the human can take in the image data at a glance, quite a large knowledge database will be needed if a machine is to interpret the scene. Of course, if some means of presenting the carcass in a highly constrained manner is available, then the vision system will have to cope with far less variability, though to some extent the situation will then be *ad hoc*, and any significant change (e.g. from beef to pig carcasses) will result in considerable effort having to be devoted to setting up the modified system. Added to this, many of the tasks that have to be carried out on carcasses—evisceration, deboning, dismembering, and so on—cannot easily be achieved by vision alone. The reason for this is that the bones are well hidden within the carcass, and vary in shape from instance to instance; in addition, the flesh itself will vary in thickness and consistency. Thus tasks such as deboning have to be carried out (and are carried out by humans) partly by touch, with all the possibilities for force feedback or compliant control that it entails. In other words biological manipulation—by human or robot—is not a simple procedure that is straightforwardly predefined in such a way that full reliance can be placed on vision. In fact, the process is more subtle: the eye or camera can take cues from what is visible, initiate hypotheses about what is invisible, and then the mechanical system can operate in a way that is partly compliant, partly orchestrated by feedback, to continue with the required function. Space does not permit further detailed analyses here of the complexities required for full effective intelligent robot control of these tasks. We merely note that meat processing in general is one of the more difficult tasks that remains to be carried out by fully automatic means (Khodabandehloo 1993a, e.g. Chapters 1 and 5).

13.5 Case study: inspection and grading of poultry parts

We now move on to a slightly less ambitious task—that of processing poultry carcasses and parts (Daley and Carey 1991; Daley *et al.* 1988, 1993). Similar statements can be made about image recognition and segmentation of whole poultry carcasses. Thresholding is a poor approach to segmentation in monochrome images, partly because of the shadows formed on the worktable, which lead to serious background clutter, and partly because of non-uniform lighting in what is after all a 3D object with a very significant range of surface orientations. However, when colour can be used, segmentation is much easier. In addition, it starts being possible to recognize some of the defects that such carcasses are subject to, namely skin tears, bruises, gall stains, feathers, discolorations, exposed flesh, broken bones, or missing parts. As in the earlier case studies in Chapter 12, HSI colour representation is used, and hue (H) is used as a colour measure. The particular value of H is that it is independent of intensity (I), so variations in lighting, or to first order, the effects of shadows, will not affect perception of bruises or other defects, so long as they really are a different colour. Once colour segmentation has been carried out, morphological operations such as erosion and median filtering can be used to eliminate some of the noise, and histogram operations can be carried out to obtain required information, e.g. on the proportion of the total area that is bruised.

Once individual items such as legs or wings have been extracted from poultry carcasses, they can be inspected and graded like many other simple shapes—as in the case of vegetables. We shall not dwell further on the work of Daley and his colleagues, as no additional image processing principles appear to be involved.

13.5.1 Tasks associated with packaging

Khodabandehloo (1993b) describes a system for automatic packaging of poultry products. The vision part of the system is needed to:

- analyse a picture of the scene
- determine the position and orientation of each object in the scene
- identify and inspect the objects
- use the resulting information to guide the robot.

The software needed to achieve this involves connected components analysis, pixel counting and grey-scale processing. Connected components analysis is required to determine the number of objects and the numbers of holes within them. Object areas and moments are then obtained, the latter being used to calculate area centroids. In this case grey-scale processing involves histogram analysis. Objects are recognized by boundary models, template matching and feature-based recognition.

A typical packaging task involves location and recognition of drumsticks, which are highly variable in shape and have to be distinguished from whole legs which consist of both drumstick and thigh. Bruises on the drumsticks must then be identified. These have grey-scale values between those of good skin (light) and cut flesh (fairly dark). Difficulties arise when trying to set thresholds to identify bruises, but it was felt that colour processing would provide a more robust solution to bruise detection. To avoid the additional cost that this would entail, stability and uniformity of lighting in a grey-scale system were found to be vital.

13.6 Concluding remarks

The case studies presented in this chapter have covered just a few of those possible in the meat, poultry and fish category, and the choice has been made somewhat didactically. Poultry, fish and prawns consist of small units: thus they fall into very much the same category as fruit and vegetables (Chapter 12). In particular, size and shape variations are important, as is the general appearance and attractiveness of the product. In addition, defects, bruising or other damage are highly relevant to its final value. Meat, on the other hand, starts off as an animal carcass, which is large and cumbersome, and grading is of less immediate importance than operations such as evisceration, dismembering and deboning. These large-scale butchery tasks are increasingly being carried out automatically or semi-automatically by robot. Unfortunately, they are complex tasks requiring considerable judgement, and vision is not able to mediate the process unaided: force feedback and non-vision sensors such as ultrasound are also vital to robot guidance during manipulation and cutting. In addition, the way vision operates in such cases is to indicate suitable reference points on a carcass, such as those provided by the outlines of the ribs or other bones, and then to determine the general trajectory for cutting: force feedback or compliance is then normally used to complete the operation, with little further help from vision. It should also be noted that the small part played by vision in advising the system of the best reference points on the carcass implies a very significant knowledge base on the part of the robot system—well beyond what has been covered in Part 1 of this monograph. All these factors mean that carcass processing is a specialized area which cannot be dealt with fully here. The reader is referred to Khodabandehloo (1993a) for further details. On the fish processing side, a useful reference is Pau and Olafsson (1991).

Other interesting work has been carried out, albeit with relatively little image processing, on oyster meat sorting (Awa *et al.* 1988; Koslav *et al*. 1989). Here the area presented by the oyster meat in the image is very quickly measured, and the weight estimated by a known relationship between these parameters, which is sufficiently accurate for the purpose. Other important aspects of the work are the need to use two crossed polarizers to eliminate the (severe) effects of specular reflection when the oysters lie in a wet environment, and the concomitant need for a high intensity light source; and the need for fast processing to cope with up to five oyster meats per second. Interested readers are referred to the original papers for further details. Further interesting applications not covered elsewhere in this chapter are:

- beef carcass evaluation by image analysis: Cross *et al*. 1983
- fluorescence of fish bones: Huss *et al*. 1984
- distinguishing between sea fish: Wagner *et al*. 1987
- defect detection in cod fillets: Bardin 1991
- lighting setup for fish inspection: Bengoetxea 1991; Piironen 1991
- fish weight estimation using structured light: Storbeck and Daan 1991
- fish processing using vision: Arnarson and Khodabandehloo 1993
- size sorting of fish: Arnarson and Ásmundsson 1993
- fish sizing and monitoring by underwater stereo: Ruff *et al*. 1995; Chan *et al*. 1999
- fish age and growth analysis by growth ring detection in otoliths (in the fish inner ear): Rodin *et al*. 1996; Benzinou *et al*. 1997.

Next, the work of Schatzki and others on X-ray inspection of meat should be mentioned as it reflects an important area that can be expected to expand further, especially with regard to the detection of tiny bones in meat and fish, which were left behind by earlier deboning and filleting operations. Schatzki *et al*. (1996) describe research aimed at the X-ray detection of foreign objects in meat. This work contrasts with that covered by the case studies of Sections 11.3–11.5 in that the meat is less textured than packs of vegetables, making detection easier. Slivers of glass, bone and wood were used for the tests, wood being the most problematic even in thicknesses of 4 mm. In addition, Schatzki and Wong (1989) did research to detect inclusions of heavy metals in meat using X-ray imaging coupled with scanning electron microscopy. This work showed the merit of X-rays for this purpose, though the techniques developed could probably only be used in case of alarm over specific toxic contaminants.

Finally, it is relevant to point to one or two early publications on standards for grading:

- EC Commission Regulation 3703/85 1986: Guidance note to the fishing industry
- US Food and Drug Administration 1988: Requirements for specific standardized fish and shellfish.

Chapter 14

System design considerations

14.1 Introduction

One of the most important limitations on the rate at which inspection systems can be produced is the lack of flexibility of existing design strategies. To some extent, this spells from our lack of understanding of the basic principles of vision upon which inspection systems might be based. In addition, there is the problem of our lack of knowledge of what goes into the design process. It is difficult enough designing a complete inspection system, including all the effort that goes into producing a cost-effective real-time hardware implementation, without having to worry at the same time whether the schema used is generic or adaptable to other products. Yet this is a crucial factor which deserves considerably more attention than hitherto.

A major factor standing in the way of progress in this area is the lack of complete case studies of inspection systems. This problem is exacerbated by industrial secrecy and profit factors, for such a system involves considerable investment by various parties and details are not lightly imparted as general scientific know-how for the benefit of all, including industrial competitors! In addition, journals generally do not permit significant space to be devoted to non-scientific factors, and certainly not to creativity aspects: they want scientific methodology and results but not what they (or the referees) regard as 'subjective chat'. Nevertheless, the latter is part of the whole design process and indeed if the situation is to develop as it should, the psychological background should be aired as part of the overall design procedure.

This chapter aims to show some of the background which is important to the design process in this area. It attempts to reveal some of the constraints to which workers are subject while designing a new inspection system, not least the tradeoffs between relevant parameters which must be

Table 14.1 Stages in the design of a typical inspection system

1. hearing about the problem
2. analysing the situation
3. looking at the data
4. testing obvious algorithms
5. realizing limitations
6. developing algorithms further
7. finding things are difficult and to some extent impossible
8. doing theory to find the source of any limitations
9. doing further tests
10. getting an improved approach
11. re-assessing the specification
12. deciding whether to go ahead
13. completing a software system
14. assessing the speed limitations
15. starting again if necessary
16. speeding up the software
17. reaching a reasonable situation
18. putting through 1000 images
19. designing a hardware implementation
20. revamping the software if necessary
21. putting through another 100,000 images
22. assessing difficulties regarding rare events
23. assessing timing problems
24. validating the final system.

optimized if the system is to be completely effective and to achieve its full potential.

14.2 Design of inspection systems—the status quo

As practised hitherto, the design of an inspection system has a number of stages, much as in the list presented in Table 14.1. However, this list is far from complete: for example, it includes no mention of lighting systems, and these are crucial to the success of any design. Nevertheless, it is a useful start, and does reveal something about the creativity aspects—by admitting that re-assessments of the efficacy of algorithms may be needed. In fact, there are necessarily one or two feedback loops, through which the efficacy can be improved systematically, again and again, until operation is adequate, or until the process is abandoned. The underlying 'process' appears to be:

```
create a basic scheme;
repeat
    improve current scheme;
    if time up then stop;
    if no further ideas then stop;
until an adequate system is obtained;
```

The "if no further ideas" clause can be fulfilled if no way is found of making the system fast enough or low enough in cost, or of high enough specification. This would account for most contingencies that could arise.

The problem with the above process is that it represents *ad hoc* rather than scientific development: it reflects a process of massaging the system or the techniques used until a satisfactory solution is obtained, irrespective of the possibility of a principled solution which builds on previous scientific methodology. Certainly, it does not help to provide, or enhance, scientific methodology. Of course, science could be introduced at any stage of the process, but the emphasis of the process is not on this—rather, on reaching a solution. There is also no guarantee that the solution that is reached is in any way optimal. Indeed, specification of the problem is not insisted upon (except to the level of 'adequacy'), and if specification and aims are absent, it becomes difficult to judge whether the success that is obtained is laudable or not. In an engineering environment we ought to be insisting on problem specification first, solution second. However, things are more complex than this—as will be seen from Table 14.2.

In the last case in Table 14.2, there are some types of fault[1] which are incredibly rare, so that only one may arise in many million cases, or, equivalently, every few weeks or months. Thus there is virtually no statistical basis for making judgements of the risk of failure, and no means of training the system so that it can learn to discriminate these faults. For these sorts of reasons, making a rigorous specification and systematically trying to meet it are difficult if not impossible, though it is still worth trying to do so. We have to try to achieve this by scientific means: if we abandon science, what else is there?

Let us return to the first of the complexities listed in Table 14.2. Although in principle it is difficult to know if there is a solution to a problem, nevertheless it is frequently possible to examine the computer image data that arise in the application, and see whether the eye can detect the faults or the foreign objects. If it can, this represents a significant step forward, as it means that it should be possible to devise a computer algorithm to do the same thing. What will then be in question will be (1) whether we are creative enough to devise such an algorithm, and (2)

---

[1] An important class of rare faults is that of foreign objects, including the hard and sometimes soft contaminants targeted by X-ray inspection systems.

Table 14.2 Complexities of the design process

- It is not always evident that there is a solution, or at least a cost-effective one.
- Specifications cannot always be made in a non-fuzzy manner.
- There is often no rigorous scientific design procedure to get from specification to solution (there is certainly no guaranteed way of achieving this optimally).
- The optimization parameters are not obvious; nor are their relative priorities clear.
- It can be quite difficult to discern whether one solution is better than another.
- Some inspection environments make it difficult to tell whether a solution is valid or not.

whether the computer can execute it sufficiently rapidly and cost-effectively to be useful.

Next, we consider whether the solution is structured or not. Can the objects, features, defects and so on be detected systematically by following some specification, such as shape, size, colour, etc.? If this is not possible, we may be in a fuzzy type of situation where statistical measures must be applied, as for textures where regularity has given over to statistical variations in brightness. In such cases, Hough transforms (for example) will give way to co-occurrence matrices or Markov processes. This is one major judgement we will probably make early on about the nature of the inspection task. Similarly, if fuzzy, ill-defined objects are to be detected, training artificial neural networks may prove more useful than overt programming.

14.3 System optimization

To proceed further, we need to examine the optimization parameters that are relevant. In fact, there are arguably rather few of these:

- sensitivity
- accuracy
- robustness
- adaptability
- reliability
- speed
- cost

Of course, these one-word parameters are as yet rather imprecise. For example, reliability can mean a multitude of things, including freedom from mechanical failures such as might arise with the camera being shaken loose, or the illumination failing; or where timing problems occur when additional image clutter results in excessive analysis time, so that the computer can no longer keep up with the real-time flow of product. Be this as it may, several of the parameters have been shown to depend upon each other—e.g. sensitivity and accuracy, cost and speed, etc. (Davies 1997e). Thus we are working in a multi-dimensional space with various constraining curves and surfaces. In the worst case, all the parameters will be interlinked, corresponding to a single constraining surface, so that adjusting one parameter forces adjustment of at least one other. There is also the possibility that the constraining surface will impose hard limits on the values of some parameters.

In fact, each algorithm will have its own constraining surface which will in general be separate from that of others. Placing all such surfaces together will create some sort of envelope surface, corresponding to the limits of what is possible with *currently available* algorithms (Figure 14.1). There will also be an envelope surface, corresponding to what is possible with *all possible* algorithms, i.e. including those which have not yet been developed. Thus there are limits imposed by human creativity and ingenuity, and of course it is not known at what stage, if ever, such limits might be overcome.

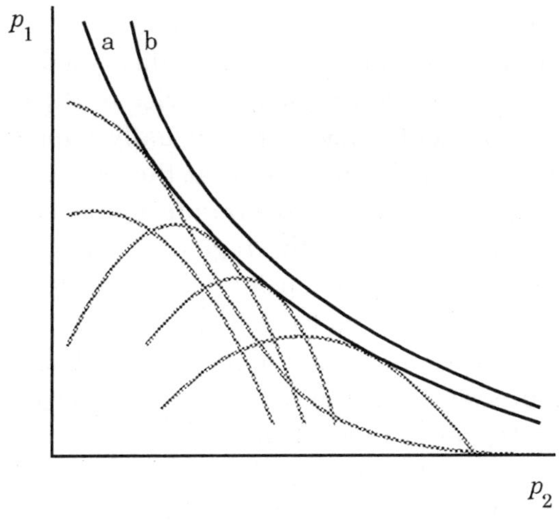

Figure 14.1 Optimization curves for a 2-parameter system. The grey curves result from individual algorithms. (a) is the envelope of the curves for all known algorithms. (b) is the limiting curve for all possible algorithms.

Returning to the constraining surface, this can be seen to provide an element of choice in the situation. Do we prefer a sensitive algorithm or a robust one, a reliable algorithm or a fast one? And so on. However, algorithms are rarely accurately describable as robust or non-robust, reliable or non-reliable, and the multi-parameter space concept with its envelope constraining surfaces clearly allows for variations in such parameters. But at the present stage of development of the subject, the problem we have is that these constraining surfaces are, with few exceptions, unknown, and the algorithms that could provide access to their ideal forms are also largely unknown; finally, even the algorithms that are available are of largely unmapped capabilities.

We can summarize the position by saying that various constraining surfaces will exist which are essentially tradeoff functions between the relevant parameters. Not only are these surfaces largely unknown, but also means are not generally known for selecting optimal working points on them (an interesting exception is that elucidated by Davies and Johnstone (1989) relating to optimization of cost/speed tradeoffs for image processing hardware). Thus there is a long way to go before the algorithm design and selection process can be made fully scientific.

14.4 The value of case studies

In the absence of scientific know-how, it may be that we are best advised to go for a case-law situation, in which we carefully define the regions where given algorithms are thought to be optimal. Thus one case study will make one claim and another case study will make another claim. If two algorithms overlap in their regions of applicability, then it may be possible to prove that the one is, in that region, superior to the other, so that the latter can be dispensed with. However, it usually happens that methods have different regions of applicability so that the one is not a subset of the other, and then both methods have to be retained independently.

More often, it happens that the regions of applicability are not so precisely known or specified. At this point let us consider inherently simple situations where objects are to be located against blank backgrounds. The only interfering factors are noise and other objects of the same type. Even here, we must appreciate that the noise may be partly Gaussian, partly impulse, partly white, partly periodic; it may also vary from one part of the image to another; and it may vary in a complex way with the grey-scale quantization. In addition, the intensity range may be wide or narrow, compressed at one end or the other, subject to exponential (as in X-ray images) or other laws, and subject to shadows or glints as well as the type of object in question. Then the objects themselves may be separate, touching, overlapping, subject to differing variabilities—in size, shape, orientation, distortion, and so on. There may be a great many of these objects or just a few; they may be small, scattered randomly, arranged in piles, appear to form a texture, and may be unsegmentable in the images. Thus even in this one 'simple' example, the potential variability is so great that it may be extremely difficult to write down all the variables and parameters of the input data so that we know exactly when a given algorithm or suite of algorithms will give an optimal solution.

Small wonder, then, that workers often seem to be re-inventing the wheel. They may in fact have worked diligently trying to get a previously available algorithm to work with their particular dataset, only to discover some major deficiency. And then, further work will have shown an appropriate route to take, with yet another algorithm which is *a priori* apparently unnecessary. So much for the adaptability that was advocated in the introduction to this chapter—the matter is far less trivial than might have been imagined.

Exploring the situation further, even low-level algorithms can be difficult to design and specify rigorously. The noise and intensity range problems referred to above still seem not to have been solved satisfactorily: we would be hard pressed to describe these problems quantitatively, and if we cannot do even this, how can we be confident of anything? In particular, if we knew that impulse and Gaussian noise were at specifiably low levels, we could confidently apply thresholding algorithms, and then proceed to perform shape analysis by skeletonization techniques. Unfortunately, this degree of confidence may rarely be justifiable: thus the odd small hole may appear in a binary shape, thereby significantly altering its skeleton. This will then radically alter the range of solutions that may realistically be adopted for image analysis[2].

14.5 The way to go

How, then, are we to develop the subject properly and make it more scientific? First, we must be optimistic, and presume that in the end knowledge will lead to solutions. Second, we must try to help things along by gathering as much precise information as possible, mostly with regard to individual cases we will be working on: for each case study we should write out working profiles in as much detail as we can, trying to get journals to publish all the available data. It can't be emphasized enough that as much description as possible should be provided defining the type of image data on which any algorithm has worked. By doing all this we will at least be

---

[2] Intrinsically, a skeleton gives a good indication of the shape of an object. Furthermore, if one noise point produces an additional loop in a skeleton, this could possibly be coped with by subsequent analysis procedures. However, significant numbers of noise points will add so much additional complexity to a skeleton that a combinatorial explosion of possible solutions will occur, with the result that the skeleton will no longer be a useful representation of the underlying shape of the object.

mapping out the constraint regions for each algorithm in the relevant parameter space in relation to the specific image data. Later, it should be possible to analyse all such mappings carefully: important lessons will then be learnt about when individual algorithms and techniques should properly be used. These ideas make some responses to the important statements in the earlier literature of the subject, which seem unfortunately to have yet had relatively little effect on most workers in the field (Jain and Binford 1991; Haralick 1992; Pavlidis 1992) (the reason for this lack of effect—especially with regard to Haralick's work—is probably that workers are very aware of the richness of the data and the algorithmic structures needed to cope rigorously with it, and have not seen quite how to proceed in the face of such complexity).

14.6 Further considerations relating to hardware accelerators

In Chapter 9, the work of Davies and Johnstone (1986, 1989) relating to the optimal deployment of hardware for performing image processing functions was outlined, and it was seen that some of the functionality should often be implemented in software on the host processor. In Chapter 11, the idea of implementing image processing functionality in software was taken considerably further: an array of DSPs programmed in a suitable software-based system was deemed to be capable of achieving the necessary throughput rates *and* to have adaptability and flexibility not normally met in hardware solutions (such as those offered by convolution boards).

At this point it is relevant to draw attention to the work of Section 10.4, which describes how pure software solutions on the host processor may sometimes provide an attractive option. First, this is even more flexible, in that special parallel processing software is not called for, so the software need not require substantial redesign if the hardware system is upgraded. Furthermore, even the array of DSPs referred to above may not be needed, so upgrades of the special software system to cope with improved varieties of such chip will not be necessary either. Of course, it may not always be possible to devise algorithms that operate sufficiently rapidly to follow real-time tasks. However, from now on, with the remarkably accelerating speeds of computers (the speedup rate is still a factor ~2 every 18 months), an increasing proportion of any required image processing functionality can reasonably be expected to match real-time requirements using software alone. This makes the implications of the work described in Section 10.4 potentially quite important.

14.7 The need for rigorous timing analysis of vision algorithms

In this subsection we examine more carefully a point which was touched upon earlier—namely the need to carry out sufficiently rigorous timing analysis of image processing tasks to ensure that the overall process will run in real time without any timing problems. In many situations it is possible for unusual image data conditions to arise which would require additional processing—thereby setting the system behind schedule: for example, this would happen if excessive noise or background clutter arose, or if more than the usual number of products appeared within a given time. A commonly used technique for eliminating this type of problem is the inclusion of a watchdog timer, connected in such a way as to curtail any current process, assuming either that the product under inspection is faulty and should be rejected, or that it is likely to be normal and that it should be accepted without further examination (another option is to deflect the product to an alternative line so that it is recycled or examined by a human operator). Watchdog timers provide crude expedient solutions, but there are signs that they will shortly be replaced by more sophisticated timing analysis methods. Môtus and Rodd are protagonists of this type of rigorous approach, which is based on Quirk analysis: see Thomas *et al.* (1995) for a basic report and Môtus and Rodd (1994) for an in-depth study.

14.8 Concluding remarks

The factors raised in the above sections, and summarized in Section 14.5, have led to the presentation of case studies at appropriate points in the last few chapters. Hopefully, they provide significant clarification of the design process and demonstrate, with a strong generic flavour, how this is carried out in the various application areas.

This chapter is not intended to be negative but rather to show up weaknesses in the design process and to indicate how the subject can be developed further for the benefit of the food industry. Perusal of the earlier chapters will quickly show that we can now achieve quite a lot even though our global design base is limited. The key to our ability to solve the problems, and to know that they are solvable, lies in the capability of the human visual system for carrying out relevant visual tasks. Note, however, that there is sound reason for replacing the human eye for these visual tasks—so that 100% tireless inspection can be carried out and so that it can be achieved consistently and reliably to known standards. At the same time, the speed of computers is maintaining its rapid rate of advance: we have seen that this could mean that there will be less need for special dedicated hardware for real-time implementation, especially if suitable algorithms such as those outlined in Section 10.4 can be developed—a very welcome factor for those developing vision systems for the food industry. On the other hand, as indicated in Section 14.7, these advances may mean that there is increased emphasis on rigorous timing analysis to ensure that inadvertent system failures cannot occur.

Chapter 15
Food processing for the Millennium

15.1 Introduction

When embarking on writing a monograph such as this, one starts off with a suitable preliminary plan based on a fair body of prior knowledge; one then assimilates all the available literature on the subject and gradually one's ideas are refined and the material starts to fall into chapters representing some sort of 'ground truth'. The situation was no different in this case. What gradually emerged was a form of presentation based very much on case-studies—known investigations relating to the application of image processing in the food industry. Indeed, the chapters in Part 2 largely represent groupings of cases which correspond quite closely to what might logically have been expected.

It is one thing to include a number of case studies, but it would have been quite another to let them dominate the whole book. If this had been done, then the monograph would have been better entitled *Image Processing in the Food Industry* rather than *Image Processing for the Food Industry*. However, it was felt to be important to retain the latter title, as it looks forward to the future rather than merely dwelling on past achievements. To justify this broader title, more talking around the subject is called for, and also considerably greater weight has to be put on what is possible in image processing, rather than what certain workers in the food industry happen to have done so far. Thus Part 1 of the monograph has taken pains to present image processing in a general way, stating clearly the concepts and their relative advantages and limitations, while particular instances of the application of image processing in Part 2 have been referred back to basic related methods in Part 1. Part 2 has also made comments on the methodology of the case studies, and has attempted to segregate such comments from those made by the original authors.

Another factor is that use of a case study should not be taken to represent complete approval by myself as author of the monograph. A case study is a case study: it is an *example* of an application design plan, but not necessarily the ultimate one. In many cases it must represent what happened for one set of investigators for one batch of foods, or indeed for a unique variety of such a batch, and the investigators may have had their hands tied in having to produce a system to strictly limited cost or other constraints. If these constraints were explicit, they have hopefully been referred to in the case study. Another factor to bear in mind is precisely when the work in question was carried out. In this subject area, events have moved fast, and have progressed significantly over the past five years, and certainly over the past decade. (Note, for example, that computers of similar cost have become more than 30 times faster over the past decade, and their memory capacity has shown similar growth.)

15.2 The range of the case studies

Looking over the various case studies in Part 2, we find that they span a very wide range of food products (Table 15.1), from the original raw product to the final processed or packed product. Here, it is interesting to compare the range of products mentioned in Table 15.1 with those listed in Tables 1.1–1.6. In this respect, the case studies included in the monograph do not seem too restricted, and are probably reasonably representative both what has been covered and what could be covered—at least in the range of foods. Looked at this way, the case studies provide existence theorems of what is possible in the application of image processing, though they do not necessarily always represent the best way of doing things: i.e. they may in some cases be lacking in reliability, adaptability, or ease of setup, and may cost more to implement than is allowable in given sectors of the industry.

The issue of adaptability is an important one, as the raw food products vary between one variety and the next, and between seasons, between countries, between regions, between fields, and also between one item and the next. A series of individual products may be perfectly acceptable, but may look different from each other. And individual sorting machines, handling robots, packing machines and inspection stations will have to operate well in spite of these variations. Equally important, the machines would ideally be able to cope with different types of produce. Why should parsnips and potatoes be sorted and graded by different machines? Why should apples, aubergines and mushrooms be inspected by different machines? At present, it is as much as system designers and image processing algorithm designers can do to make a machine which can optimally cope with a single commodity, but there is much scope for introducing more flexibility into these systems to make them less specialized and thus offset capital costs better. Potentially, this can be achieved by two main means: (1) use of suitable training procedures, and (2) use of systems which incorporate some form of knowledge base.

Table 15.1 Foods, commodities and products covered in the case studies and references

fruit

- apples
- bananas
- citrus fruits
- dates
- oranges
- peaches
- prunes
- tomatoes

vegetables

- asparagus
- bell peppers
- cabbages
- carrots
- cauliflowers
- mushrooms
- potatoes
- processed vegetables
- soybeans
- sweet potatoes

cereals and nuts

- cereal grains
- maize and soybean
- nuts
- peanuts
- pistachio nuts
- wheat kernels

animal products

- fish
- meat carcasses
- oysters
- poultry
- prawns

baked and processed products

- biscuits
- bread and loaves
- cherries (when included)
- chocolate bars
- chocolate biscuits
- cream biscuits
- Jaffacakes

packaging

- canned products
- frozen food packs
- metal and glass containers
- packaged foods

15.3 The cost of vision hardware

We have already referred to the rapidly reducing cost of computer hardware. Ten or so years ago, there was a tendency to cut algorithms to the bone in order to ensure that grading and sorting systems could be introduced into the industry. This meant that the systems then being introduced were sometimes less powerful than might have been desired. Similarly, the cost of hardware accelerators to implement vision algorithms in real time was excessive and this meant that in certain cases it was not economic to include them on production lines, even if all the necessary scientific investigation had been done and otherwise viable solutions had been found. The time is now approaching when the extra hardware needed to achieve real-time operation will be minimal, partly because the host computer will be so powerful that it will be able to cope with all relevant (if carefully chosen) algorithms almost unaided. Hence speed of processing and hardware cost of the vision system will no longer be deciding parameters. The remaining costs will be: scientific analysis costs and costs of software development; costs of the mechanical handling machinery (though the latter should be minimal if included when the production line is installed). In this milieu, there will essentially be no excuse for not including 'due diligence' vision inspection capability on all foods and food products.

15.4 The potential range of applications of vision

The next question is how widely image processing can be applied within the food industry. First, it should be noted that image processing is an enabling technology. It permits things to be done, and robot actions to be taken—often in the way that mimics human activity. It is interesting to explore for a minute the overall food production scenario. Raw food products are grown; they are then processed in various ways; they are graded, sorted, packaged and inspected. Specifically fruit, vegetable and grain sources and products are subjected to the following processes:

- pruning
- razing
- sowing
- growing
- weeding
- spraying
- tending
- harvesting
- threshing
- milling
- crushing
- stoning

(the last of these applying to certain types of fruit such as cherries and olives) while meat, fish and poultry products are subjected to the following processes:

- feeding/grazing
- tending
- killing
- evisceration
- deboning

- filleting
- dismembering
- skinning
- plucking.

Finally, food preparation is subject to:

- baking
- blanching
- coating
- cooking

and many more such processes. These are followed by:

- inspecting
- grading
- sorting
- canning
- packing
- marketing.

Note also that marketing may occur in many stages.

Over all these processes, considerable value is often added to the products, which can even become technological masterpieces of decorative engineering. Many of these processes are readily carried out by robots or equivalent stationary devices, often with the help of a machine vision system. The higher the value added, the more likely it is to be done indoors in a highly constrained, strictly controlled environment. In such cases, vision is straightforward to engineer. In general, vision is difficult to manage in unconstrained situations—especially outdoors where most of the raw products come from. The lists of processes to which fruit, vegetables and grain are subjected (see above) include some apparently simple but actually exceedingly difficult tasks—pruning, tending, harvesting—of which the last is often the most difficult of all. Similarly, the list of processes to which animal products are subject starts with some of the most difficult ones to manage. The point in both cases is that the unconstrained and less well-defined tasks are the most difficult. To emulate a human operator's performance demands a considerable database of knowledge: e.g. a fruit picker must anticipate where to find the fruit, recognize it, locate the gripping points, notice that it is not yet sufficiently ripe, and be capable of logging relevant information on the state of the trees, presence of insects, mould, etc. If the fruit is partly obscured by leaves, branches and so on, and has a hitherto unseen combination of colours, the vision system is going to be hard pressed to cope well with the situation, though valiant attempts have been made to do so in specific cases. Similar comments apply to behavioural analysis of animal movements made without apparent effort by stockmen, but not yet achieved fully automatically by machine.

15.5 Prognosis

All this indicates that there is some way to go before vision can be used to mediate all mechanization on the farm and before full use is made of the potential of vision to help machines emulate the human worker—whether for labour saving or for optimal management in view of the exodus from the country and lack of experienced workers on every country site. Remote health checks of plants and animals are yet to come, but during the first decade of the Millennium they should have arrived, with the help of computer vision systems.

We are living in a rapidly changing world: food products in general have increasing value added; quality is the concern of farmer and consumer alike; there is increasing concern about health and safety; due diligence is here; litigation is a growth area; and no sector of the food industry remains untouched by rapidly advancing technology. These trends seem certain to continue unabated into the 21st century.

References

Abutaleb, A.S. (1989) Automatic thresholding of gray-level pictures using two-dimensional entropy. *Comput. Vision Graph. Image Process.* **47**, 22–32

Ade, F. (1983) Characterization of texture by 'eigenfilters'. *Signal Process.* **5**, no. 5, 451–457

Analog Devices (1990) *ADSP 2101 Users' Manual (Architecture).* Norwood, Mass., USA

Arcelli, C. and di Baja, G.S. (1985) A width-independent fast-thinning algorithm. *IEEE Trans. Pattern Anal. Mach. Intell.* **7**, 463–474

Arcelli, C., Cordella, L.P. and Levialdi, S. (1975) Parallel thinning of binary pictures. *Electron. Lett.* **11**, 148–149

Arcelli, C., Cordella, L.P. and Levialdi, S. (1981) From local maxima to connected skeletons. *IEEE Trans. Pattern Anal. Mach. Intell.* **3**,134–143

Arnarson, H. and Ásmundsson, M. (1993) Computer vision in food handling and sorting. Chapter 4.1 in Chen, C.H., Pau, L.F. and Wang, P.S.P. (eds.) *Handbook of Pattern Recognition and Computer Vision*, World Scientific Pub. Co., pp. 697–718

Arnarson, H. and Khodabandehloo, K. (1993) Fish processing using computer vision and robots. In Khodabandehloo (1993a), pp. 26–47

Awa, T.W., Byler, R.K. and Diehl, K.C. (1988) Development of an inexpensive oyster meats grader. ASAE Paper no. 88-3539

Bailey, D.G. and Hodgson, R.M. (1988) VIPS – a digital image processing algorithm development environment. *Image Vision Comput.* **6**, no. 3, 176–182

Ball, G.H. and Hall, D.J. (1966) ISODATA, an iterative method of multivariate data analysis and pattern classification. *Proc. IEEE Int. Communications Conf.*, Philadelphia, Digest of Techn. Papers II, pp. 116–117

Ballard, D.H. (1981) Generalizing the Hough transform to detect arbitrary shapes. *Pattern Recogn.* **13**, 111–122

Bangham, J.A. and Marshall, S. (1998) Image and signal processing with mathematical morphology. *IEE Electronics and Commun. Eng. Journal* **10**, no. 3, 117–128

Bardin, E. (1991) Laboratory work using the DT 2851 frame grabber: application of computer vision to inspection of fish. In Pau and Olafsson (1991), pp. 225–244

Batchelor, B.G. (1993) Automated inspection of bread and loaves. *Machine Vision Applications, Architectures and Systems Integration II* (7–9 Sept.), *SPIE* **2064**, 124–134

Batchelor, B.G., Hill, D.A. and Hodgson, D.C. (eds.) (1985) *Automated Visual Inspection*. IFS (Publications) Ltd., Bedford, UK/North Holland, Amsterdam

Bengoetxea, K. (1991) Lighting setup in the automatic detection of ventral skin and blood spots in cod fish fillets. In Pau and Olafsson (1991), pp. 121–150

Benzinou, A., Troadec, H., Le Bihan, J., Rodin, V., de Pontual, H. and Tisseau, J. (1997) The locally deformable B-bubble model: an application to growth ring detection on fish otoliths. *Proc. 10th Scand. Conf. on Image Analysis*, Lappeenranta, Finland (9–11 June), pp. 181–187

Bertrand, D., Guernevé, C., Marion, D., Devaux, M.-F. and Robert, P. (1992) Description of the textural appearance of bread crumb by video image analysis. *Cereal Chem*. **69**, no. 3, 257–261

Bolles, R.C. and Cain, R.A. (1982) Recognizing and locating partially visible objects: the local-feature-focus method. *Int. J. Robot. Res*. **1**, no. 3, 57–82

Brady, J.M. and Wang, H. (1992) Vision for mobile robots. *Phil. Trans. R. Soc. (London)* **B337**, 341–350

Brink, A.D. (1992) Thresholding of digital images using two-dimensional entropies. *Pattern Recogn*. **25**, 803–808

Brivot, R. and Marchant, J.A. (1996) Segmentation of plants and weeds for a precision crop protection robot using infrared images. *IEE Proc. Vision Image Signal Process*. **143**, no. 2, 118–124

Cambier, J.L. and Pasiak, D.C. (1986) On-line X-ray inspection of packaged foods. *Proc. SME Conf. on Vision*, Detroit, Section 1, pp. 1–14

Casady, W.W. and Paulsen, M.R. (1989) An automated kernel positioning device for computer vision analysis of grain. *Trans. Amer. Soc. Agric. Eng*. **32**, no. 5, 1821–1826

Casasent, D., Talukder, A., Lee, H.-W., Keagy, P.M., and Schatzki, T.F. (1997) Segmentation and classification for X-ray agricultural product inspection. *Proc. 5th Int. Symp. on Fruit Nut Vegetable Engineering*, UC Davis (Sept.)

Chambers, J. (1997) Revealing invertebrate pests in food. In *Crop protection and food quality: meeting customer needs, Proc. BCPC/ANPP Conf.*, Canterbury, Kent (16–19 Sept.), pp. 363–370

Chambers, J., and Ridgway, C. (1996) Rapid detection of contaminants in cereals. In *Near Infrared Spectroscopy: The Future Waves*, Davies,

A.M.C. and Williams, P. (eds.), NIR Publications, Chichester, UK, pp. 484–489

Chan, D., Hockaday, S., Tillett, R.D. and Ross, L.G. (1999) A trainable n-tuple pattern classifier and its application for monitoring fish underwater. *Proc. 7th IEE Conf. on Image Processing and its Applications*, Manchester (12–14 July), IEE Conf. Publication. no. 465, pp. 255–259

Chan, J.P. and Batchelor, B.G. (1993) Machine Vision for the Food Industry. Ch. 4 in Pinder, A.C. and Godfrey, G. (1993), pp. 58–101

Chan, J.P., Batchelor, B.G., Harris, I.P. and Perry, S.J. (1990) Intelligent visual inspection of food products. *Machine Vision Systems Integration in Industry*, Boston (Nov.), *SPIE* **1386**, 171–179

Chen, S., Fon, D., Wu, C., Leu, K., Tien, B. and Chang, W. (1992) Electro-optical citrus sorter. ASAE Paper no. 92-3520, St. Joseph, Mich.

Cho, S.I., Saito, H. and Ozawa, S. (1997) Analytical solution of shape from shading problem. *Proc. 8th Brit. Mach. Vision Assoc. Conf.*, Univ. of Essex (8–11 Sept.), pp. 51–59

Chow, C.K. and Kaneko, T. (1972) Automatic boundary detection of the left ventricle from cineangiograms. *Comput. Biomed. Res.* **5**, 388–410

Chtioui, Y., Bertrand, D., Dattée, Y. and Devaux, M.-F. (1996) Identification of seeds by colour imaging: comparison of discriminant analysis and artificial neural network. *J. Sci. Food Agric.* **71**, 433–441

Churchill, D.B., Bilsland, D.M. and Cooper, T.M. (1992) Comparison of machine vision with human measurement of seed dimensions. *Transactions of the ASAE* **35**, no. 1, 61–64

Conners, R.W. and Harlow, C.A. (1980) A theoretical comparison of texture algorithms. *IEEE Trans. Pattern Anal. Mach. Intell.* **2**, no. 3, 204–222

Cootes, T.F., Taylor, C.J., Cooper, D.H. and Graham, J. (1992) Training models of shape from sets of examples. *Proc. 3rd Brit. Mach. Vision Assoc. Conf.*, Leeds, (22–24 Sept.) pp. 9–18

Cronshaw, A.J. (1982) Automatic chocolate decoration by robot vision. *Proc. 12th Int. Symp. Ind. Robots (ISIR)*

Cross, H.R., Gilliland, D.A., Durland, P.R. and Seideman, S. (1983) Beef carcass evaluation by use of an video image analysis system. *J. Animal Sci.* **57**, no. 4, 908–917

Cybenko, G. (1988) Continuous valued neural networks with two hidden layers are sufficient. Techn. Report, Dept. of Comput. Sci., Tufts Univ., Medford, MA

D'Agostino, S. (1990) Machine vision based length distribution analysis for the food industry. *Proc. Food Processing Automation Conf.*, Lexington, Kentucky, pp. 72–77

Daley, W.D.R. and Carey, R.A. (1991) Color machine vision for defect detection (algorithms and techniques). *Proc. Int. Robots and Vision Automation Conf.*, Detroit (Oct.), pp. 5-29–5-42

Daley, W.D.R., Millet, R., Soulakos, C. and Thompson, J.C. (1988) A novel application: machine vision inspection, grading and parts identification of chicken parts. *Proc. SME Robots 12 and Vision '88*, Detroit (June), pp. 13-33–13-49

Daley, W.D.R., Wyvill, J.C., Thompson, J.C., Holcombe, W.D. and McMurray, G.V. (1993) Robotics and the poultry processing industry. In Khodabandehloo (1993a), pp. 48–69

Davenel, A., Guizard, C.H., Labarre, T. and Sevila, F. (1988) Automatic detection of surface defects on fruit by using a vision system. *J. Agric. Eng. Res.* **41**, no. 1, 1–9

Davies, E.R. (1984a) Design of cost-effective systems for the inspection of certain foodproducts during manufacture. In Pugh, A. (ed.), *Proc. 4th Conf. on Robot Vision and Sensory Controls*, London (9–11 Oct.), pp. 437–446

Davies, E.R. (1984b) Circularity–a new principle underlying the design of accurate edge orientation operators. *Image Vision Comput.* **2**,134–142

Davies, E.R. (1985) Radial histograms as an aid in the inspection of circular objects. *IEE Proc. D* **132**, no. 4, Special Issue on *Robotics*, 158–163

Davies, E.R. (1986) Image space transforms for detecting straight edges in industrial images. *Pattern Recogn. Lett.* **4**, no. 3, 185–192

Davies, E.R. (1987) A high speed algorithm for circular object location. *Pattern Recogn. Lett.* **6**, no. 5, 323–333

Davies, E.R. (1988a) A hybrid sequential-parallel approach to accurate circle centre location. *Pattern Recogn. Lett.* **7**, no. 5, 279–290

Davies, E.R. (1988b) Tradeoffs between speed and accuracy in two-stage template matching. *Signal Process.* **15**, no. 4, 351–363

Davies, E.R. (1988c) A modified Hough scheme for general circle location. *Pattern Recogn. Lett.* **7**, 37–43

Davies, E.R. (1988d) Median-based methods of corner detection. In: Kittler, J. (ed.) *Proceedings of the 4th BPRA International Conference on Pattern Recognition, Cambridge (28–30 March)*. Lecture Notes in Computer Science, Vol. 301. Springer-Verlag, Heidelberg, pp. 360–369

Davies, E.R. (1989a) Edge location shifts produced by median filters: theoretical bounds and experimental results. *Signal Process.* **16**, no. 2, 83–96

Davies, E.R. (1989b) Finding ellipses using the generalised Hough transform. *Pattern Recogn. Lett.* **9**, 87–96

Davies, E.R. (1991a) Alternative to abstract graph matching for locating objects from their salient features. *Image Vision Comput.* **9**, no. 4, 252–261

Davies, E.R. (1991b) The minimal match graph and its use to speed identification of maximal cliques. *Signal Process.* **22**, no. 3, 329–343

Davies, E.R. (1992a) Locating objects from their point features using an optimised Hough-like accumulation technique. *Pattern Recogn. Lett.* **13**, no. 2, 113–121

Davies, E.R. (1992b) Simple two-stage method for the accurate location of Hough transform peaks. *IEE Proc. E* **139**, no. 3, 242–248

Davies, E.R. (1993) *Electronics, Noise and Signal Recovery*, Academic Press, London

Davies, E.R. (1995) Machine vision in manufacturing – what are the real problems? *Proc. 2nd Int. Conf. on Mechatronics and Machine Vision in Practice*, Hong Kong (12–14 Sept.), pp. 15–24

Davies, E.R. (1997a) Designing efficient line segment detectors with high orientation accuracy. *Proc. 6th IEE Int. Conf. on Image Processing and its Applications*, Dublin (14–17 July), IEE Conf. Publication no. 443, pp. 636–640

Davies, E.R. (1997b) Lower bound on the processing required to locate objects in digital images. *Electronics Lett.* **33**, no. 21, 1773–1774

Davies, E.R. (1997c) Principles and design graphs for obtaining uniform illumination in automated visual inspection. *Proc. 6th IEE Int. Conf. on Image Processing and its Applications*, Dublin (14–17 July), IEE Conf. Publication no. 443, pp. 161–165

Davies, E.R. (1997d) Vectorial strategy for designing line segment detectors with high orientation accuracy. *Electronics Lett.* **33**, no. 21, 1775–1777

Davies, E.R. (1997e) *Machine Vision: Theory, Algorithms, Practicalities*, Academic Press, London (2nd edition)

Davies, E.R. (1997f) Algorithms for inspection: constraints, tradeoffs and the design process. IEE Digest no. 1997/041, *Colloquium on Industrial Inspection*, IEE (10 Feb.), pp. 6/1–5

Davies, E.R. (1998) Rapid location of convex objects in digital images. *Proc. European Signal and Image Processing Conf.* (EUSIPCO'98), Rhodes, Greece, 8–11 Sept., pp. 589–592

Davies, E.R. (1999) Chord bisection strategy for fast ellipse location. *Electronics Lett.* **35**, no. 4, 703–705

Davies, E.R. and Johnstone, A.I.C. (1986) Engineering trade-offs in the design of a real-time system for the visual inspection of small products. *Proc. 4th Conf. on UK Research in Advanced Manufacture*, London (10–11 Dec.), IMechE Conf. Publications 1986–12, pp. 15–22

Davies, E.R. and Johnstone, A.I.C. (1989) Methodology for optimising cost/speed tradeoffs in real-time inspection hardware. *IEE Proc. E* **136**, no. 1, 62–69

Davies, E.R. and Plummer, A.P.N. (1981) Thinning algorithms: a critique and a new methodology. *Pattern Recogn.* **14**, 53–63.

Davies, E.R., Bateman, M., Chambers, J. and Ridgway, C. (1998a) Hybrid non-linear filters for locating speckled contaminants in grain. IEE

Digest no. 1998/284, *Colloquium on Non-Linear Signal and Image Processing*, IEE (22 May), pp. 12/1–5

Davies, E.R., Bateman, M., Mason, D.R., Chambers, J. and Ridgway, C. (1999) Detecting insects and other dark line features using isotropic masks. *Proc. 7th IEE Int. Conf. on Image Processing and its Applications*, Manchester (13–15 July), IEE Conf. Publication no. 465, pp. 225–229

Davies, E.R., Mason, D.R., Bateman, M., Chambers, J. and Ridgway, C. (1998b) Linear feature detectors and their application to cereal inspection. *Proc. European Signal and Image Processing Conf.* (EUSIPCO'98), Rhodes, Greece, 8–11 Sept., pp. 2561–2564

Davies, E.R., Patel, D. and Johnstone, A.I.C. (1995) Crucial issues in the design of a real-time contaminant detection system for food products. *Real-Time Imaging* **1**, no. 6, 397–407

Davis, P.F. and Tillett, R.D. (1994) Image analysis with artificial neurons to select cutting planes for the micropropagation of syngoniums. *Comput. Electron. Agric.* **10**, 189–202

Delagnes, P., Benois, J. and Barba, D. (1995) Active contours approach to object tracking in image sequences with complex background. *Pattern Recogn. Lett.* **16**, no. 2, 171–178

Delwiche, M.J., Tang, S. and Thompson, J.F. (1990a) Prune defect detection by line-scan imaging. *Trans. Amer. Soc. Agric. Eng.* **33**, no. 3, 950–954

Delwiche, M.J., Tang, S. and Thompson, J.F. (1990b) A high speed multiprocessor inspection system for dried prunes. *Proc. Food Processing Automation Conf.*, Lexington, Kentucky, pp. 281–288

Dennis, T.J., Clark, L.J. and Gunawardena, C.A. (1988) Computer controlled sizing and quality control system for agricultural produce. *Proc. 2nd Int. Conf. Computers in Agric.*, Univ. Florida, Inst. Food and Agric. Sci.

Devijver, P.A. and Kittler, J. (1982) *Pattern Recognition: a Statistical Approach*. Prentice-Hall, Englewood Cliffs, NJ

Dougherty, E.R. (1992) *An Introduction to Morphological Image Processing*. SPIE Press, Bellingham, MA

Dougherty, E.R. and Giardina, C.R. (1988) Morphology on umbra matrices. *Int. J. Pattern Recogn. and Artif. Intell.* **2**, 367–385

Dowell, F.E. (1992) Automatic control of a peanut grade sample inspection system. *Food Control* **3**, no. 2, 105–108

Duda, R.O. and Hart, P.E. (1973) *Pattern Classification and Scene Analysis*. Wiley, New York

EC Commission Regulation 3703/85 (1986) *Guidance note to the fishing industry*

Edmonds, J.M. and Davies, E.R. (1991) High-speed processor for realtime visual inspection. *Microprocessors and Microsystems* **15**, no. 1, 11–19

Faugeras, O.D. (1993) *Three-Dimensional Computer Vision – a Geometric Viewpoint*. MIT Press, Cambridge, MA

Fischler, M.A. and Bolles, R.C. (1981) Random sample consensus: a paradigm for model fitting with applications to image analysis and automated cartography. *Commun. ACM* **24**, no. 6, 381–395

Forgy, E.W. (1965) Cluster analysis of multivariate data: efficiency versus interpretability of classification. Biometrics **21**, 768–769

Gonzalez, R.C. and Woods, R.E. (1992) *Digital Image Processing*. Addison Wesley, Reading, Ma

Graf, G.L., Rehkugler, G.E. and Millier, W.F. (1981) Automatic detection of surface flaws on apples using digital image processing. ASAE Paper, no. 81-3537

Graves, M., Batchelor, B.G. and Palmer, S. (1994) 3D X-ray inspection of food products. *Applications of Digital Image Process.* **17**, *SPIE* **2298**, 248–259

Graves, M., Smith, A. and Batchelor, B.G. (1998) Approaches to foreign body detection in foods. *Trends in Food Science and Technology* **9**, no. 1, 21–27

Greenhill, D. and Davies, E.R. (1994) Relative effectiveness of neural networks for image noise suppression. In Gelsema, E.S. and Kanal, L.N. (eds.) *Pattern Recognition in Practice IV*, Elsevier Science B.V., pp. 367–378

Hamey, L.G.C. and Yeh, J.C.-H. (1996). Segmentation of bake images by a self-organising map. *Proc. Image Segmentation Workshop*, Sydney, Australia (Dec.), pp. 65–68

Hampel, F.R., Ronchetti, E.M., Rousseeuw, P.J. and Stahel, W.A. (1986) *Robust Statistics. The Approach Based on Influence Functions*, Wiley, New York

Hannah, I., Patel, D. and Davies, E.R. (1995) The use of variance and entropic thresholding methods for image segmentation. *Pattern Recogn.* **28**, no. 8. 1135–1143

Haralick, R.M. (1979) Statistical and structural approaches to texture. *Proc. IEEE* **67**, no. 5, 786–804

Haralick, R.M. (1980) Edge and region analysis for digital image data. *Comput. Graph. Image Process.* **12**, 60–73

Haralick, R.M. (1992) Performance characterization in image analysis: Thinning, a case in point. *Pattern Recogn. Lett.* **13**, 5–12

Haralick, R.M., Shanmugam, K. and Dinstein, I. (1973) Textural features for image classification. *IEEE Trans. Systems Man Cybern.* **3**, no. 6, 610–621

Haralick, R.M., Sternberg, S.R. and Zhuang, X. (1987) Image analysis using mathematical morphology. *IEEE Trans. Pattern Anal. Mach. Intell.* **9**, no. 4, 532–550

Harwood, D., Subbarao, M. and Davis, L.S. (1985) Texture classification by local rank correlation. *Comput. Vision Graph. Image Process.* **32**, 404–411

Haykin, S. (1994) *Neural Networks*. Macmillan College Pub. Co., New York

Heijmans, H. (1994) *Morphological Operators*. Academic Press, New York

Heinemann, P.H., Hughes, R., Morrow, C.T., Sommer, H.J. III, Beelman, R.B. and Wuest, P.J. (1994) Grading of mushrooms using a machine vision system. *Trans. Amer. Soc. Agric. Eng.* **37**, no. 5, 1671–1677

Heinemann, P.H., Pathare, N.P. and Morrow, C.T. (1996) An automated inspection station for machine-vision grading of potatoes. *Machine Vision Applications* **9**, no. 1, 14–19

Heinemann, P.H., Varghese, Z.A., Morrow, C.T., Sommer, H.J. III and Crassweller, R.M. (1995) Machine vision inspection of 'Golden Delicious' apples. *Appl. Eng. Agric.* **11**, no. 6, 901–906

Hertz, J., Krogh, A. and Palmer, R.G. (1991) *Introduction to the Theory of Neural Computation*. Addison-Wesley, Reading, MA

Hildreth, E.C. (1984) *Measurement of Visual Motion*. MIT Press, Cambridge, MA

Hodgson, R.M. and McNeill, S.J. (1986) The design of image processing systems for real time inspection applications. *Proc. 2nd IEE Int. Conf. on Image Processing and its Applications*, IEE Conf. Publication no. 265, pp. 126–129

Horaud, R., Conio, B., Leboulleux, O. and Lacolle, B. (1989) An analytic solution for the perspective 4-point problem. *Comput. Vision Graph. Image Process.* **47**, 33–44

Horn, B.K.P. (1986) *Robot Vision*. MIT Press, Cambridge, MA

Howarth, M.S. and Searcy, S.W. (1992) Inspection of fresh market carrots by machine vision. *Proc. Conf. on Food Processing Automation*, ASAE, St. Joseph, Mich., pp. 106–111

Hsiao, J.Y. and Sawchuk, A.A. (1989) Supervised textured image segmentation using feature smoothing and probabilistic relaxation techniques. *IEEE Trans. Pattern Anal. Mach. Intell.* **11**, no. 12, 1279–1292

Hsiao, J.Y. and Sawchuk, A.A. (1990) Unsupervised textured image segmentation using feature smoothing and probabilistic relaxation techniques. *Comput. Vision Graph. Image Process.* **48**, 1–21

Hush, D.R. and Horne, B.G. (1993) Progress in supervised neural networks. *IEEE Signal Processing Magazine* **10**, no. 1, 8–39

Huss, H.H., Sigsgaard, P. and Jensen, S.A. (1984) Fluorescence of fish bones. *J. Food Protection* **48**, 393–396

Image Inspection (1994) *Centurion Data Sheet*. Epsom, Surrey, UK

Jain, A.K. and Dubes, R.C. (1988) *Algorithms for Clustering Data*. Prentice-Hall, Englewood Cliffs, NJ

Jain, R.C. and Binford, T.O. (1991) Ignorance, myopia and naiveté in computer vision systems. CVGIP: Image Understanding **53**, 112–117

Johnson, M. (1985) Automation in citrus sorting and packing. Agri-Mation I Conf. and Expos, Chicago, Il., pp. 63–68

Kapur, J.N., Sahoo, P.K. and Wong, A.K.C. (1985) A new method for gray-level picture thresholding using the entropy of the histogram. *Comput. Vision Graph. Image Process.* **29**, 273–285

Kass, M. and Witkin, A. (1987) Analyzing oriented patterns. *Comput. Vision Graph. Image Process.* **37**, no. 3, 362–385

Kass, M., Witkin, A. and Terzopoulos, D. (1988) Snakes: active contour models. *Int. J. Comput. Vision* **1**, 321–331

Kassler, M. (1990) Robotics and prawn-handling. *Robotica* **8**, no. 4, 299–301

Kassler, M., Corke, P.I. and Wong, P.C. (1991) Automatic grading and packing of prawns. *Proc. Int. Robots and Vision Automation Conf.*, Detroit (Oct.), pp. 9-27–9-32

Keagy, P.M. and Schatzki, T.F. (1993) Machine recognition of weevil damage in wheat radiographs. *Cereal Chem.* **70**, no. 6, 696–700

Keagy, P.M., Parvin, B. and Schatzki, T.F. (1995) Machine recognition of navel orange worm damage in x-ray images of pistachio nuts. *Optics in Agric., Forestry and Biological, SPIE* **2345**, 192–203

Keagy, P.M., Parvin, B. and Schatzki, T.F. (1996) Machine recognition of navel orange worm damage in X-ray images of pistachio nuts. *Lebens.-Wiss. u.-Technol.* **29**, 140–145

Keefe, P.D. (1992) A dedicated wheat grain image analyser. *Plant Varieties and Seeds* **5**, 27–33

Khodabandehloo, K. (1993b) Robotic packaging of poultry products. In Khodabandehloo (1993a), pp. 70–97

Khodabandehloo, K. (ed.) (1993a) *Robotics in Meat, Fish and Poultry Processing*, Blackie, Glasgow

Kittler, J., Illingworth, J. and Föglein, J. (1985) Threshold selection based on a simple image statistic. *Comput. Vision Graph. Image Process.* **30**, 125–147

Koslav, M.B., Byler, R.K. and Diehl, K.C. Jr. (1989) Machine vision based oyster meat sorter. ASAE Paper no. 89-7518, pp. 1–17

Krueger, K. (1991) Can robots handle prawn? J. Australian Fisheries

Kuhn, E.D., Ambrose, J.T. and Unrath, C.R. (1982) A measurement technique for delicious apple shape. Hort. Sci. **17**, 785–787

Kuo, F.F. (1966) *Network Analysis and Synthesis*. (2nd edition) Wiley, New York

Laws, K.I. (1979) Texture energy measures. *Proc. Image Understanding Workshop*, Nov., pp. 47–51

Laws, K.I. (1980) Rapid texture identification. *Proc. SPIE Conf. on Image Processing for Missile Guidance* **238**, San Diego, Calif., 28 July – 1 Aug., pp. 376–380

Liao, K., Cavalieri, R.P. and Pitts, M.J. (1990) Hausdorff dimensional analysis and digital imaging based quality inspection. *Trans. Amer. Soc. Agric. Eng.* **33**, no. 1, 298–304

Liao, K., Paulsen, M.R. and Reid, J.F. (1994) Real-time detection of colour and surface defects of maize kernels using machine vision. *J. Agric. Eng. Res.* **59**, 263–271

Lippmann, R.P. (1987) An introduction to computing with neural nets. *IEEE Acoustics, Speech, Signal Process. Magazine* **4**, no. 2, 4–22

MacQueen, J.B. (1967) Some methods for classification and analysis of multivariate observations. *Proc. 5th Berkeley Symp. on Math. Stat. and Prob.*, I, pp. 281–297

Maeda, H. (1987) Grade and colour sorter through image processing. *Proc. Int. Symp. Agric. Mach. and Int. Co-op in High Tech. Era*, Univ. Tokyo, pp. 350–356

MAFF (1973) *EEC standards for fresh apples and pears*. Publication of Ministry of Agriculture, Fisheries and Food, UK

Majumdar, S., Jayas, D.S. and Bulley, N.R. (1997) Classification of cereal grains using machine vision, part 1: morphological features; part 2: color features; part 3: textural features; part 4: color, textural and morphological features. *Proc. ASAE Annual International Meeting*, Minneapolis (10–14 August)

Marchant, J.A. (1996) Tracking of row structure in three crops using image analysis. *Comput. Electron. Agric.* **15**, 161–179

Marchant, J.A. and Brivot, R. (1995) Real-time tracking of plant rows using a Hough transform. *Real-Time Imaging* **1**, no. 5, 363–371

Marchant, J.A. and Onyango, C.M. (1995) Fitting grey level point distribution models to animals in scenes. *Image Vision Comput.* **13**, no. 1, 3–12

Marchant, J.A. and Tillett, R.D. (1994) Software and transputer system design for high speed grading of agricultural produce. *Mechatronics* **4**, no. 3, 281–293

Marchant, J.A., Onyango, C.M. and Street, M.J. (1988) High speed sorting of potatoes using computer vision. ASAE Paper no. 88-3540

Marchant, J.A., Onyango, C.M. and Street, M.J. (1990) Computer vision for potato inspection without singulation. *Comput. Electron. Agric.* **4**, 235–244

Marchant, J.A., Tillett, R.D. and Brivot, R. (1998) Real-time segmentation of plants and weeds. *Real-Time Imaging* **4**, 243–253

Marslin, R.F., Sullivan, G.D. and Baker, K.D. (1991) Kalman filters in constrained model based tracking. *Proc. 2nd Brit. Mach. Vision Assoc. Conf.*, Glasgow (23–26 Sept.), pp. 371–374

McClure, J.E. and Morrow, C.T. (1987) Computer vision sorting of potatoes. ASAE Paper no. 87-6501

McFarlane, N.J.B. and Schofield, C.P. (1995) Segmentation and tracking of piglets in images. *Machine Vision Applications* **8**, no. 3, 187–193

Meer, P., Mintz, D., Rosenfeld, A. and Kim, D.Y. (1991) Robust regression methods for computer vision: a review. *Int. J. Comput. Vision* **6**, no. 1, 59–70

Miller, B.K. and Delwiche, M.J. (1989a) A colour vision system for peach grading. *Trans. Amer. Soc. Agric. Eng.* **32**, no. 4, 1484–1490

Miller, B.K. and Delwiche, M.J. (1989b) Peach defect detection with machine vision. ASAE Paper no. 89-6019

Môtus, L. and Rodd, M.G. (1994) *Timing Analysis of Real-Time Software.* Elsevier, Oxford

Mundy, J.L. and Zisserman, A. (eds.) (1992) *Geometric Invariance in Computer Vision.* MIT Press, Cambridge, MA

Nagel, H.-H. (1983) Displacement vectors derived from second-order intensity variations in image sequences. *Comput. Vision Graph. Image Process.* **21**, 85–117

Nagel, H.-H. (1986) Image sequences–ten (octal) years–from phenomenology towards a theoretical foundation. *Proc. 8th Int. Conf. on Pattern Recogn., Paris (27–31 October)*, pp. 1174–1185

Naim, M.M., Chan, J.P. and Huneiti, A.M. (1994) Quantifying the learning curve of a vision inspection system. *Proc. IEE 4th Int. Conf. on Advanced Factory Automation*, York (Nov.), IEE Conf. Publication no. 398, pp. 509–516

Newman, T.S. and Jain, A.K. (1995) A survey of automated visual inspection. *Comput. Vision Image Understanding* **61**, no. 2, 231–262

Novini, A.R. (1990) Fundamentals of machine vision inspection in metal container and glass manufacturing. *Proc. SME Vision Conf. '90*, Detroit (Nov.), pp. 14-1–14-11

Oja, E. (1989) Neural networks, principal components, and subspaces. *Int. J. Neural Systems* **1**, 61–68

Onyango, C.M. and Marchant, J.A. (1996) Modelling grey level surfaces using three-dimensional point distribution models. *Image Vision Comput.* **14**, 733–739

Otsu, N. (1979) A threshold selection method from gray-level histograms. *IEEE Trans. Systems Man Cybern.* **9**, 62–66

Paler, K., Föglein, J., Illingworth, J. and Kittler, J. (1984) Local ordered grey levels as an aid to corner detection. *Pattern Recogn.* **17**, 535–543

Patel, D., Davies, E.R. and Hannah, I. (1995) Towards a breakthrough in the detection of contaminants in food products. *Sensor Review* **15**, no. 2, 27–28

Patel, D., Davies, E.R. and Hannah, I. (1996) The use of convolution operators for detecting contaminants in food images. *Pattern Recogn.* **29**, no.6, 1019–1029

Patel, D., Hannah, I. and Davies, E.R. (1994) Texture analysis for foreign object detection using a single layer neural network. *Proc. IEEE Int. Conf. on Neural Networks*, Florida (28 June–2 July), Vol. **VII**, pp. 4265–4268

Pau, L.F. and Olafsson, R. (eds.) (1991) *Fish Quality Control by Computer Vision*, Marcel Dekker, New York

Paulsen, M.R., Wigger, W.D., Litchfield, J.B. and Sinclair, J.B. (1989) Computer image analyses for detection of maize and soybean kernel quality factors. *J. Agric. Eng. Res.* **43**, no. 2, 93–101

Pavlidis, T. (1992) Why progress in machine vision is so slow. *Pattern Recogn. Lett.* **13**, 221–225

Pearson, T.C. (1996) Machine vision system for automated detection of stained pistachio nuts. *Lebens.-Wiss. u.-Technol.* **29**, no. 3, 203–209

Pearson, T.C. and Slaughter, D.C. (1996) Machine vision detection of early split pistachio nuts. *Trans. Amer. Soc. Agric. Eng.* **39**, no. 3, 1203–1207

Penman, D., Olsson, O. and Beach, D. (1992) Automatic x-ray inspection of canned products for foreign material. *Machine Vision Applications, Architectures and Systems Integration*, *SPIE* **1823**, 342–347

Peterson, P.E.H. and Krutz, G.W. (1992) Automatic identification of weed seeds by colour machine vision. *Seed Science and Seed Technology* **20**, no. 2, 193–208

Pfaltz, J.L. and Rosenfeld, A. (1967) Computer representation of planar regions by their skeletons. *Comm. ACM* **10**, 119–125

Pietikäinen, M., Rosenfeld, A. and Davis, L.S. (1983) Experiments with texture classification using averages of local pattern matches. *IEEE Trans. Systems Man Cybern.* **13**, no. 3, 421–426

Piironen, T. (1991) Evaluation of illumination methods for machine vision applications in the food industry. In Pau and Olafsson (1991), pp. 97–120

Pinder, A.C. and Godfrey, G. (eds.) (1993) *Food Process Monitoring Systems*, Blackie and Son, Glasgow

Press, W.H., Teukolsky, S.A., Vetterling, W.T. and Flannery, B.P. (1992) *Numerical Recipes in C. The Art of Scientific Computing.* Cambridge Univ. Press, Cambridge (2nd edition)

Pun, T. (1980) A new method for grey-level picture thresholding using the entropy of the histogram. *Signal Process.* **2**, 223–237

Reed, T.R. and Du Buf, J.M.H. (1993) A review of recent texture segmentation and feature extraction techniques. *CVGIP: Image Understanding* **57**, 359–372

Rehkugler, G.E. and Throop, J.A. (1985) Apple sorting with machine vision. ASAE Paper no. 85-3543

Rehkugler, G.E. and Throop, J.A. (1989) Image processing algorithm for apple defect detection. *Trans. Amer. Soc. Agric. Eng.* **32**, no. 1, 267–272

Ridgway, C. and Chambers, J. (1996) Detection of external and internal insect infestation in wheat by near-infrared reflectance spectroscopy. *J. Sci. Food Agric.* **71**, 251–264

Ridgway, C. and Chambers, J. (1998) Detection of insects inside wheat kernels by NIR imaging. *J. Near Infrared Spectroscopy* **6**, 115–119

Rigney, M.P., Brusewitz, G.H. and Hranzler, G.A. (1992) Asparagus-defect inspection with machine vision. ASAE Paper no. 92-403, St. Joseph, Mich.

Rodin, V., Troadec, H., de Pontual, H., Benzinou, A., Tisseau, J. and Le Bihan, J. (1996) Growth ring detection on fish otoliths by a graph construction. *Proc. IEEE Int. Conf. on Image Processing*, Lausanne, Switzerland (16–19 Sept.), pp. 685–688

Rogers, D.E., Day, D.D. and Olewnik, M.C. (1995) Development of an objective crumb-grain measurement. *Cereal Foods World* **40**, no. 7, 498–501

Rosenblatt, F. (1958) The perceptron: a probabilistic model for information storage and organisation in the brain. *Psychol. Review* **65**, 386–408

Rosenblatt, F. (1962) *Principles of Neurodynamics*. Spartan, New York

Rosenfeld, A. (1970) Connectivity in digital pictures. *J. ACM* **17**, 146–160

Rosenfeld, A. (1979) *Picture Languages: Formal Models for Picture Recognition*. Academic Press, New York

Rosenfeld, A. and Pfaltz, J.L. (1966) Sequential operations in digital picture processing. *J. ACM* **13**, 471–494

Rosenfeld, A. and Pfaltz, J.L. (1968) Distance functions on digital pictures. *Pattern Recogn.* **1**, 33–61

Ruff, B.P., Marchant, J.A. and Frost, A.R. (1995) Fish sizing and monitoring using a stereo image analysis system applied to fish farming. *Aquacultural Engineering* **14**, 155–173

Rumelhart, D.E., Hinton, G.E. and Williams, R.J. (1986) Learning internal representations by error propagation. In Rumelhart, D.E. and McClelland, J.L. (eds.), *Parallel Distributed Processing: Explorations in the Microstructure of Cognition*, The MIT Press, Cambridge, Mass., pp. 318–362

Sahoo, P.K., Soltani, S., Wong, A.K.C. and Chen, Y.C. (1988) A survey of thresholding techniques. *Comput. Vision Graph. Image Process.* **41**, 233–260

Sanchiz, J.M., Pla, F., Marchant, J.A. and Brivot, R. (1996) Structure from motion techniques applied to crop field mapping. *Image Vision Comput.* **14**, 353–363

Sanger, T.D. (1989) Optimal unsupervised learning in a single-layer linear feedforward neural network. *Neural Networks* **2**, 459–473

Sangwine, S.J. and Horne, R.E.N. (1998) *The Colour Image Processing Handbook*. Chapman and Hall, London

Sarkar, N. and Wolfe, R.R. (1985a) Feature extraction techniques for sorting tomatoes by computer vision. *Trans. Amer. Soc. Agric. Eng.* **28**, no. 3, 970–974

Sarkar, N. and Wolfe, R.R. (1985b) Computer-vision-based system for quality separation of fresh market tomatoes. *Trans. Amer. Soc. Agric. Eng.* **28**, 1714–1718

Sarkar, N. and Wolfe, R.R. (1986) Computer vision based system for quality separation of fresh market tomatoes. ASAE Paper no. 86-6539, St. Joseph, Mich.

Schatzki, T.F. and Wong, R.Y. (1989) Detection of submilligram inclusions of heavy metals in processed foods. *Food Technology* **43**, no. 11, 72–76

Schatzki, T.F., Haff, R.P., Young, R., Can, I., Le, L.-C. and Toyofuku, N. (1997) Defect detection in apples by means of X-ray imaging. *Trans. Amer. Soc. Agric. Eng.* **40**, no. 5, 1407–1415

Schatzki, T.F., Young, R., Haff, R.P., Eye, J.G. and Wright, G.R. (1996) Visual detection of particulates in x-ray images of processed meat products. *Opt. Eng.* **35**, no. 8., 2286–2291

Serra, J. (1982) *Image Analysis and Mathematical Morphology*. Academic Press, New York

Shatadal, P., Jayas, D.S., Hehn, J.L. and Bulley, N.R. (1995) Seed classification using machine vision. *Canadian Agricultural Engineering* **37**, no. 3, 163–167

Shearer, S.A. and Payne, F.A. (1990) Machine vision sorting of bell peppers. *Proc. Food Processing Automation Conf.*, Lexington, Kentucky, pp. 289–300

Shirai, Y. (1987) *Three-dimensional Computer Vision*. Springer-Verlag, Berlin

Storbeck, F. and Daan, B. (1991) Weight estimation of flatfish by means of structured light and image analysis. *Fisheries Research* **11**, 99–108

Strachan, N.J.C. (1993) Recognition of fish species by colour and shape. *Image Vision Comput.* **11**, no. 1, 2–10

Strachan, N.J.C. (1994) Sea trials of a computer vision based fish species sorting and size grading machine. *Mechatronics* **4**, no. 8, 773–783

Strachan, N.J.C., Nesvadba, P. and Allen, A.R. (1990) Fish species recognition by shape analysis of images. *Pattern Recogn.* **23**, no. 5, 539–544

Tan, T.N., Sullivan, G.D. and Baker, K.D. (1994) Recognizing objects on the ground-plane. *Image Vision Comput.* **12**, no. 3, 164–172

Tang, S., Delwiche, M.J. and Thompson, J.F. (1989) Development and evaluation of a defect sorter for prunes. ASAE Paper no. 90-6039, St. Joseph, Mich.

Tao, Y., Heinemann, P.H., Varghese, Z.A., Morrow, C.T. and Sommer, H.J. III (1995a) Machine vision for color inspection of potatoes and apples. *Trans. Amer. Soc. Agric. Eng.* **38**, no. 5, 1555–1561

Tao, Y., Morrow, C.T. and Heinemann, P.H. (1990) Automated machine vision inspection of potatoes. ASAE Paper no. 90-3531

Tao, Y., Morrow, C.T., Heinemann, P.H. and Sommer, H.J. III (1995b) Fourier-based separation technique for shape grading of potatoes using machine vision. *Trans. Amer. Soc. Agric. Eng.* **38**, no. 3, 949–957

Taylor, J. and Coombes, S. (1993) Learning higher order correlations. *Neural Networks* **6**, 423–427

Taylor, R.W. and Rehkugler, G.E. (1985) Development of a system for automated detection of apple bruises. *Proc. SME Conf. on Vision*, Detroit (Mar.), Section 6, pp. 55–65

Thomas, A.D.H., Rodd, M.G., Holt, J.D. and Neill, C.J. (1995) Real-time industrial visual inspection: a review. *Real-Time Imaging* **1**, no. 2, 139–158

Thomson, W.H. and Pomeranz, Y. (1991) Classification of wheat kernels using three-dimensional image analysis. *Cereal Chem.* **68**, no. 4, 357–361

Throop, J.A. and Aneshanesley, D.J. (1993) Investigation of texture analysis features of apple bruises. ASAE Paper no. 93-3527, St. Joseph, Mich.

Tillett, R.D. (1990) Vision-guided planting of dissected microplants. *J. Agric. Eng. Res.* **46**, 197–205

Tillett, R.D. and Batchelor, B.G. (1991) An algorithm for locating mushrooms in a growing bed. *Comput. Electron. Agric.* **6**, 191–200

Tillett, R.D. and McFarlane, N.J.B. (1991) Understanding plant images. *Proc. 1st Int. Seminar on Use of On-Machine Vision Systems for the Agricultural and Bio-Industries*, CEMAGREF Montpellier, France (3–6 Sept.), pp. 133–140

Tillett, R.D., Brown, F.R., McFarlane, N.J.B., Onyango, C.M., Davis, P.F. and Marchant, J.A. (1990) Image-guided robotics for the automation of micropropagation. *Proc. IEEE Int. Workshop on Intell. Robots and Systems,* IROS'90, pp. 265–270

Tillett, R.D., Onyango, C.M. and Marchant, J.A. (1997) Using model-based image processing to track animal movements. *Comput. Electron. Agric.* **17**, 249–261

Tollner, E.W., Affeldt, H.A. Jr., Brown, G.K., Chen, P., Galili, N., Haugh, C.G., Notea, A., Sarig, Y., Schatzki, T.F., Shmulivich, I. and Zion, B. (1993) Nondestructive detection of interior voids, foreign inclusions and pests

Tsuji, S. and Matsumoto, F. (1978) Detection of ellipses by a modified Hough transform. *IEEE Trans. Comput.* **27**, 777–781

Turney, J.L., Mudge, T.N. and Volz, R.A. (1985) Recognizing partially occluded parts. *IEEE Trans. Pattern Anal. Mach. Intell.* **7**, 410–421

Unser, M. (1986) Local linear transforms for texture measurements. *Signal Process.* **11**, 61–79

Unser, M. and Eden, M. (1989) Multiresolution feature extraction and selection for texture segmentation. *IEEE Trans. Pattern Anal. Mach. Intell.* **11**, no. 7, 717–728

Unser, M. and Eden, M. (1990) Nonlinear operators for improving texture segmentation based on features extracted by spatial filtering. *IEEE Trans. Systems Man Cybern.* **20**, no. 4, 804–815

US Food and Drug Administration (1988) *Requirements for specific standardized fish and shellfish.* Code of Federal Regulations, Title 21, Part 161.130–161.140

USDA (1966) *US standards for grades of mushrooms.* US Department of Agriculture—Consumer and Marketing Service, Washington, DC

USDA (1978) *Apples—market inspection instructions.* US Department of Agriculture—Fruit and Vegetable Division, Fresh Products Branch

USDA (1991) *US standards for grades of potatoes.* US Department of Agriculture—Agricultural Marketing Service, Washington, DC

Van Gool, L., Dewaele, P. and Oosterlinck, A. (1985) Survey: Texture analysis anno 1983. *Comput. Vision Graph. Image Process.* **29**, 336–357

Varghese, Z.A., Morrow, C.T. and Heinemann, P.H. (1991) Automated inspection of Golden Delicious apples using colour computer vision. ASAE Paper no. 91-7002

Vistnes, R. (1989) Texture models and image measures for texture discrimination. *Int. J. Comput. Vision* **3**, 313–336

Wagner, G.G. (1987) Combining X-ray imaging and machine vision. *Optics, Illumination and Image Sensing for Machine Vision II, SPIE* **850**, 45–53

Wagner, H., Schrnidt, U. and Rudeck, J.H. (1987) Distinction between species of sea fish. *Lebensmittel Industrie* **34**, 20–23

Whitworth, M.B. and Alava, J.M. (1999) The imaging and measurement of bubbles in bread doughs. In Campbell *et al.* (eds.) *Bubbles in Food*, Eagan Press, St. Paul, pp. 221–231

Whitworth, M.B., Evers, T.D. and Brock, C.J. (1997) On-line measurement of bran in flour by image analysis. In Campbell *et al.* (eds.) *Cereals: Novel Uses and Processes*, Plenum Press, New York, pp. 125–131

Wigger, W.D., Paulsen, M.R., Litchfield, J.B. and Sinclair, J.B. (1988) Classification of fungal-damaged soybeans using color-image processing. ASAE Paper no. 88-3053, St. Joseph, Mich.

Wolfe, R.R. and Swaminathan, M. (1987) Determining orientation and shape of bell peppers by machine vision. *Trans. Amer. Soc. Agric. Eng.* **30**, no. 6, 1853–1856

Wright, M.E., Sistler, F.E. and Watson, R.M. (1984) Measuring sweet potato size and shape with a computer. *Louisiana Agric.* **28**, no. 1, 12–13

Wu, Q.M. and Rodd, M.G. (1991) Fast boundary extraction for industrial inspection. *Pattern Recogn. Lett.* **12**, no. 8, 483–489

Wulfsohn, D., Sarig, Y. and Algazi, R. (1989) Preliminary investigation to identify parameters for sorting dates by image processing. ASAE Paper no. 89-6610

Yang, Q. (1992) The potential for applying machine vision to defect detection in fruit and vegetable grading. *Agric. Eng.* **47**, no. 3, 74–79

Yang, Q. (1994) An approach to apple surface feature detection by machine vision. *Comput. Electron. Agric.* **11**, 249–264

Yang, Q. and Marchant, J.A. (1996) Accurate blemish detection with active contour models. *Comput. Electron. Agric.* **14**, 77–89

Yeh, J.C.-H., Hamey, L.G.C., Westcott, T. and Sung, S.K.Y. (1995). Colour bake inspection system using hybrid artificial neural networks. *Proc. IEEE Int. Conf. Neural Networks*, Perth, Australia (Nov.), pp. 37–42

Yuen, H.K., Illingworth, J. and Kittler, J. (1988) Ellipse detection using the Hough transform. *Proc. 4th Alvey Vision Conf., Manchester* (*31 August–2 September*), pp. 265–271

Zayas, I.Y. and Flinn, P.W. (1998) Detection of insects in bulk wheat samples with machine vision. *Trans. Amer. Soc. Agric. Eng.* **41**, no. 3, 883–888

Zayas, I.Y. and Steele, J.L. (1990) Image analysis applications for grain science. *Optics in Agric., SPIE* **1379**, 151–161

Zhang, J. and Modestino, J.W. (1990) A model-fitting approach to cluster validation with application to stochastic model-based image segmentation. *IEEE Trans. Pattern Anal. Mach. Intell.* **12**, no. 10, 1009–1017

Zhuang, X. and Haralick, R.M. (1986) Morphological structuring element decomposition. *Comput. Vision Graph. Image Process.* **35**, 370–382

Author Index

Subject Index

THE UNIVERSITY OF READING LIBRARY